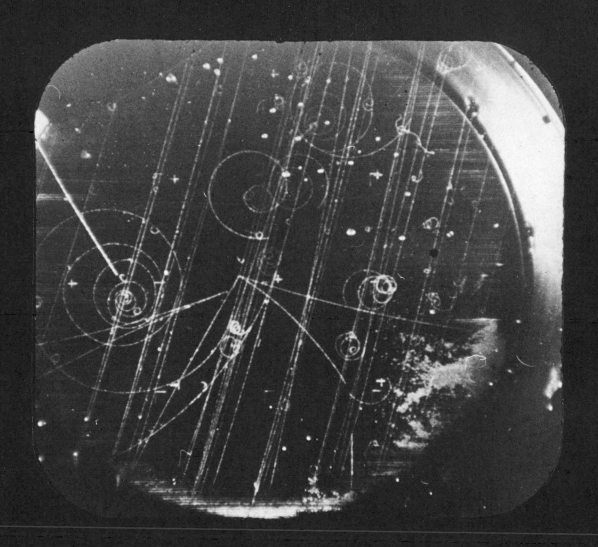

FABRIC OF THE UNIVERSE

FABRIC OF THE UNIVERSE

Denis Postle

Crown Publishers, Inc., New York

Acknowledgements

This book is dedicated to the several dozen people who have so generously contributed time, energy and material to its creation. I hope they find that it adequately reflects the quality and quantity of attention that they have all brought to it (and earlier, to the film from which it was born).

I am most especially indebted to Dr. John Charap, Reader in Theoretical Physics at Queen Mary College, London University. The patience, generosity and insight of his teaching appear at all levels of the book, not only in its content but also in its presentation. While his presence is indelibly impressed on the matters of physics which we present, he is not however responsible for the opinions and speculations expressed herein. These have been assembled by the author, who must bear sole responsibility for any errors or inadequacies which may reveal themselves.

The open-minded encouragement over a period of several years of Edwin Shaw, Director of Public Information at CERN, provided essential practical support without which the project would certainly have withered. Nic Knowland has been a source of advice and insight on the various aspects of the Eastern teachings touched upon. I'd also like to express my thanks to everyone who read and commented on the manuscript, especially Johannah Freudenberg, Dr. I. Halliday, Mal Hawley, Dr. E. Holiday, Gina Martin, and Jeremy Newson, and to my wife for her many, many hours of typing and, more important, the down-to-earth common sense she has never failed to provide.

Lastly, this book is also dedicated to Maharishi Mahesh Yogi for the gift of the technique of transcendental meditation without which it would certainly never have been written.

The publishers thank Granada Publishing Limited for permission to quote the three poems by E. E. Cummings on pages 202 and 203.

Endpapers: stereo pairs of bubble chamber photographs showing the tracks of charged subatomic particles in a hydrogen bubble chamber. To view them, place a small mirror against the right side of your nose, holding the mirror about 1 ft above the centre fold. Then adjust reflected picture (right) until it merges with picture (left)

First published in the USA in 1976
by Crown Publishers, Inc
ISBN 0-517-52726X
Library of Congress Catalog Card Number 76-2340

Printed and bound in Hong Kong

Contents

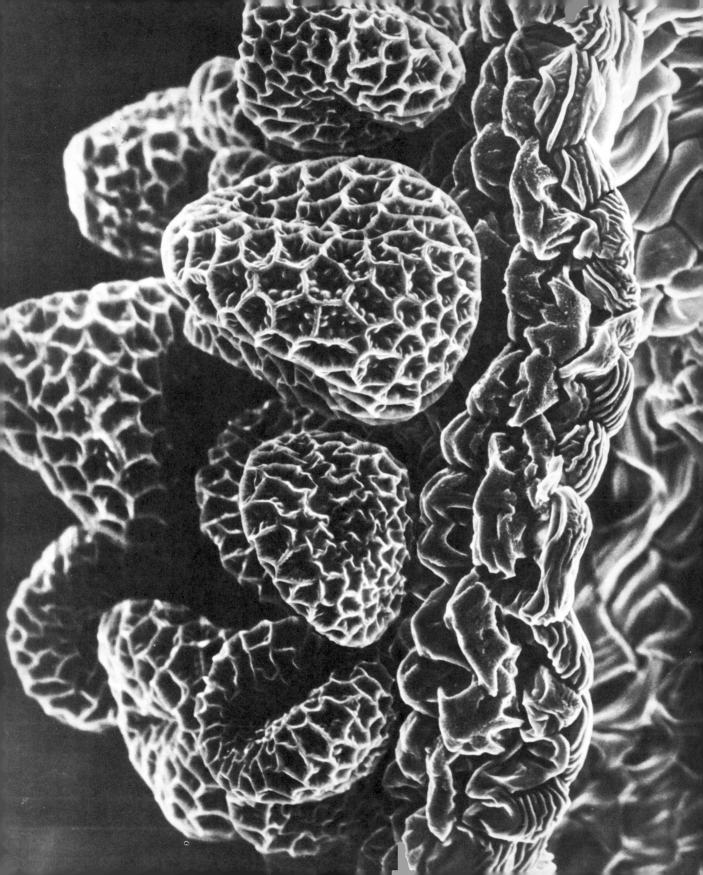

Introduction

This book has one or two things to say about what is normal on the cosmic scale. By normal is meant that which is usual, typical, regular—by cosmic scale is meant everything, everywhere. Everything that is.

Much of what we know about what is normal on the cosmic scale, and therefore most of the material in this book, is drawn from that branch of science which seeks to describe the fundamental laws governing the structure and behaviour of matter—particle physics.

But particle physics is a remote and esoteric area of science, so specialized in its techniques, and of such difficulty, that it is poorly understood even by other scientists. Is access possible? Taken head on, with its 'Regge poles', 'probability waves' and 'charmed quarks', physics *is* very nearly impenetrable. But, seen in the light of another, apparently separate discipline which also claims to know something about the cosmic scale—Eastern philosophy, it looks less complex.

This then is a book in which teachings from the Eastern philosophic tradition, embracing Hinduism, Taoism and Buddhism, are used to illuminate some of the discoveries of Western science, especially those of particle physics. It is not a book about religion, nor is it one which suggests that science is a religion; neither is it a science teaching text (few scientists are named in it). It does not follow the usual historical discovery by discovery presentation. It makes no claim to being a comprehensive survey of the current state of physics, exciting though that may be to physicists close to the action.

The subject of the book is not particle physics itself, but what particle physics *reveals* about the world.

The aim throughout has been to de-mystify particle physics and to open up points of contact between it and the non-scientist. The book is as far as possible a practical one, full of recipes for action, things to make and do. Though, as with a normal cook-book, it's not assumed that everyone will try every recipe.

Chapters One and Two begin by looking at what we take to be the normal on the *human* scale – 'How's life been treating you?', 'How's it going?', 'Comment ça va?', 'Could be worse?', 'A bit heavy?', 'Something less than perfect?'. The Eastern tradition teaches that we are perfect, that the universe is perfect, at least in the sense that we are, and the universe is, *complete*. The suggestion being that we already have everything we need. But, and it's a very big but, because we insist on ignoring the perfection, the completeness, we are often confused in ourselves, and from time to time we even experience, dare we say it, suffering. The first two chapters pose the questions: 'Is this suffering necessary? Is there an alternative? To what extent are we in contact with reality?'.

Chapters Three and Four introduce us to the scientific view of what constitutes reality—to what science, through its many technical extensions of touch, sight, taste, hearing, smell, thinking, memory and intelligence, has to show us. We begin to see revealed something of the world beyond the senses, and to sense the limitations of the senses. We begin also to see that, as the range and power of these scientific extensions are broadened and refined, so they point first to greater complexity and diversity in matter, and then, as mind gets to work, to simplicity within the diversity. (As, for example, the microscope at first revealed thousands of different kinds of bacteria, which later could be seen to belong to rather fewer different types.)

In Chapters Five to Ten we move on to see what is revealed by the most expensive and elaborate extensions of the senses yet built—particle accelerators, which are in effect town-sized microscopes. Elsewhere in science, only telescopes compare with them for size and cost.

Here again, as this vast refinement of traditional microscopy has looked in on nature's inner structure, there has been, particularly since 1960, a vast extension of apparent diversity. But now, in the mid-seventies, the beginnings of simplicity are starting to emerge. This knowledge that is emerging from particle physics is at once the core of this book and the reason for its existence, because the simplicity that is being revealed is so deep, in fact *so* simple, that it may have become a signpost. Pointing to itself. Pointing to unity.

Not just simplicity among the multitudinous diversity of the material substance of the world, but a unity containing the diversity. This is where particle physics appears to converge with the Eastern teachings. The tradition of the East has taught for centuries that, however it may seem, the true state of the universe is one of total interconnectedness, total interpenetration, total interdependence of all things and all life everywhere. Simply, unity. Particle physics speaks more tentatively, but physicists will freely admit to thinking/believing that the universe has been constructed in the only possible way, that the universe we have is the only universe that can exist. Of course, if the universe were *not* consistent, there could be no science as we know it. As Albert Einstein is reputed to have said, 'The most incomprehensible thing about the universe is that it *is* comprehensible'.

This book suggests that the accumulation of physics theory and experiment do indeed constitute a signpost pointing also to unity.

The consequences of this, if it is a fact and not just an assertion or a belief, are obviously far reaching. Several hundred years ago in the Copernican revolution, science disproved medieval man's most basic belief—that he and the earth were at the centre of the universe. Are we now to understand that science has demolished contemporary man's most basic belief, that each of us is a separate independent being, able freely to occupy the centre of our own private world? That, if we could only realize it, 'I', 'Me' and 'mine' are now dead or have no meaning? If the unity does exist, why do we not experience it? Why do we experience so much frustration, anxiety and stress? Chapters

Eleven and Twelve look at these and other questions which seem to bring the discoveries of particle physics even closer to the Eastern tradition, and in so doing bring them both closer to the practical facts of daily life.

At their most subtle level, what East and West have in common is a joint concern with consciousness itself. Physics has this century become aware of the crucial role of the consciousness of the experimenter, the observer who may be conducting and experimenting—i.e. the extent to which his attitude to the experiment may influence its outcome, and more important, the way that the decision to *do* the experiment changes what is then looked at.

Eastern teaching goes even further in this direction and says that what we *can know* depends on our consciousness (more exactly it is suggested that knowledge is structured in consciousness).

With Chapter Twelve we come to see how it is that our experience of thought, feeling and sensations overshadows the nature of consciousness itself. Overshadows our essential identity, our sense of who we are. The book concludes with the suggestion of some practical steps we can take to make personal experiments in, and gain experience of, this consciousness. Which is our true nature. Which is us.

1. Air, Metal, Stone and Bodies

Every day we live a lie. A huge untruth. Each of us takes part in the conspiracy. It's the falsehood of normality. Most of us are embarrassed by emphatic expressions of belief that carry a label like Catholic, or capitalist, and yet we never make a movement or a decision in our lives without invoking a great catalogue of beliefs and habitual attitudes. All of these beliefs hinge on what we take to be normal. The purpose of this book is to undermine these beliefs in you, and to draw aside the veils that obscure a deeper, more subtle normality.

You are probably reading this page sitting down. If not, let us imagine that you are sitting, reading in a room at home. You are surrounded by glass, metal, upholstery, wood, other people perhaps, newspapers, air, electricity, leather, etc. In addition there may be words, sounds, colours, smells, textures. For the moment we are speaking only of the physical world. If we were to add the inside mental world as well, we would have thoughts, anxieties, dreams, feelings and sensations too. A veritable rag bag of items, each one separately labelled. Or rather, labelled *separate*. Because the overriding constant in what we take to be normal is separateness of thought, action and object.

Nine men, nine minds—how many realities? New Yorkers watch the construction of an office block

Things seem separate in our minds. Are they separate in reality? Sit up, and look at some object nearby, the window frame perhaps; anything will do. Now look at it closely as if you were going to have to describe it in detail to someone. Now leave the looking for a moment and try to be aware of your weight sitting on the seat. Feel the pressure being exerted, feel the edge of the seat pressing into the flesh. Now, something else, listen for the sounds of the room where you are. The tick of the clock, the water running through the pipes, the distant traffic, the creak of leather as your shoes move on the floor. Got it? O.K., now one last thing to complete the demonstration. Take in all three at once. Listening, sitting and seeing. This is a practical book, try it. Now what happens? The chances are that you find you can't take in all three at once. When we want to know something, we concentrate our attention on one thing at a time, and yet all three things are *there simultaneously*. The effect of this kind of unconscious concentration is to make us less aware of the rest of our surroundings. We grasp at pieces of the world and tend to miss it's wholeness.

This example also serves to demonstrate something of the way that the

mind, operating through habits of which we are unaware, 'makes' the world we perceive. We become convinced that, i.e. we believe that, we live in a world of different things, different people. But is it? Is this difference, this separation, real? Or is it a dream, an apparition—a result of our ignorance of the real, i.e. whole and undivided state of the world? Perhaps our experience of the chair we're sitting on, the wall facing us and the sound of the traffic outside is only separate because we've come to believe in them as separate.

This is the lie that we inhabit, which devours us, consuming our energy, poisoning our bodies through a deluge of anxieties, phobias, dreams, and desires. If we can come to realize what this lie is made of and how it operates we can begin to be free of it. Because it inhibits effective action and blinds and deafens us to what is there.

It would be merely humourous if our sense of separation did not lead us into such an intense belief in the normality of our experience that we think this is all there is. But is it? If you doubt that this is a practical question, that it is any more than a flickering intellectual flame seeking to illuminate only the writer's ego, step outside your house and look at the stars. Do it now. O.K., what did you see? The chances are that you saw either a blue sky, or *maybe* a cloudy sky. But maybe there were stars. Yet you know that bright points of light in the sky *are* stars. How do you know? You remembered? You know

Electric lighting illuminates the foreground of our lives, the pavements, rooms and roads, but as its brightness reflects back from the atmosphere it also cuts us off from a clear view of the sky—obscuring what might otherwise be a constant reminder of our place in the scale of things

there are stars even though you can't see them. What I am trying to convey is that the piece of action you just undertook, if you did get up and go out, was largely your mind playing with the thought or idea of 'star'. You 'know' the stars are there even when you can't see them, just as you know that microbes are there even though you can't see them.

Perhaps by now you begin to see that this is the starting point of our game together. That if we put aside our prejudices, it is possible to see that what we take to be real and normal in the world is in fact information originating at the working surface of the eyes, ears, skin, tongue, or nose, being from there transmitted to the brain where old information is updated against it. And through this process, the brain or 'the mind' *makes* our room, or the tube train. Or as you read, maybe makes **frmo tehse wrdos no thsi pgae** a convincing idea. This process of the mind 'making', of the mind as active agent, is at the root of the lie we inhabit.

There is a story from the Eastern tradition which may help us see our situation a little better. It tells of a monkey locked in an empty house, a house with five windows representing the five senses. The monkey runs restlessly from one window to the next. He is imprisoned. He doesn't know how he got into this prison, so he assumes he has always been there. He does not realize that the house is a fantasy originating in his mind, superimposed on the real jungle. He feels the texture of the walls and gives names to parts of it, decides he likes some of it and is frightened of other parts. He tries to escape but fails. He then feels dejected, helpless; eventually he relaxes and lets his mind wander; he dreams. But some of his dreams terrify him: so our monkey begins to wish for something good, something beautiful and affectionate.

He now dreams of a life of freedom and ease. But having experienced this dream of bliss and delight he begins to worry that someone may thwart him in his attempts to make it come true. He starts to suffer from paranoia and defends himself against it by building up his pride. About now our monkey becomes reconciled to the mundane nature of his situation; instead of switching back and forth between pride and jealousy he begins to feel comfortable, at home in the world. He becomes adjusted to leading a regular life, doing things ordinarily. But the monkey senses this is a bit dull. There is a lack of spontaneity. The more his dreams solidify the more stupid and heavy he feels. It's all a terrible nightmare. But it's so terrible it can't be real. He begins to hate himself a little; his mental jigsaw puzzle erupts into hostility, his thought patterns become irregular and uncertain . . . insanity looms.

If this model of the mind at work (it's a caricature, of course) makes no contact with your experience, you may be one of the lucky few who always knew the prison was an artifact of the mind. Only you can know that. If you feel it has nothing to do with you at all perhaps you should reflect on the fact that mental illness afflicts about one in eight of the UK population at some time in their lives, and that the commonest causes for admissions to many big hospitals are drug overdose, coronary failure, stomach ulcers, heart failure and bronchitis, all stress-related illnesses. Could this be the mind at work too?

If it does connect with any part of your experience, then fine; but remember mind may be the cause of bondage, but it's also the means of liberation.

Opposite: the mind 'making'. Is this a picture of a man enjoying his food or a picture of food enjoying a man?

Overleaf: the mind 'making'. Look at these faces: what sort of qualities do they suggest? Cruelty? Kindness? Fear? Friendship?

14

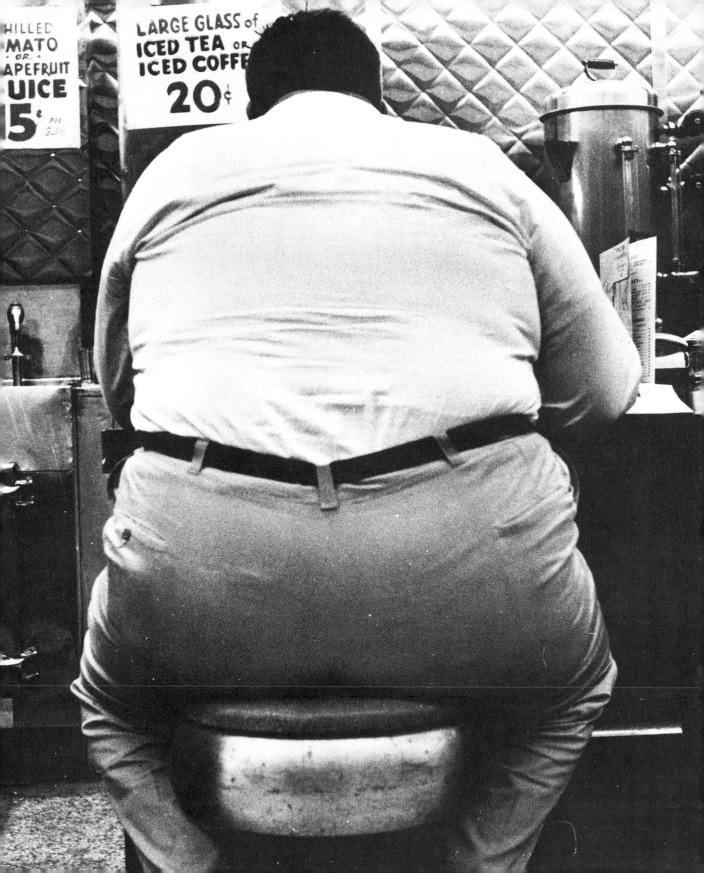

The first problem we must face is the problem of the normal. The normal for us here as we read these words and the other normal which is the subject of this book, the normal on the cosmic scale.

PLOUCH R^{D.}

The
SUPER CINEMA
WHITTLESEY
PRESENTS:

MON. JULY 27   PETER SELLERS
FOR 3 DAYS.   ERIC SYKES
              BERNARD MILES

HEAVENS ABOVE
Also: HAPPY ANNIVERSARY

ELVIS PRESLEY   THURS. JULY 30
                FOR 3 DAYS.
LOVE IN LAS VEGAS
JEFFREY HUNTER   PANAVISION.
                        METROCOLOR
GOLD FOR THE CAESARS

S.V.
4' 10"

Left: the mind 'making':
some men build up towards the sky to draw closer to
something unattainable. Other men reach up
to rid themselves of unsaleable refuse. For some, Heaven is
above; for others, the heavens are a dump. Yet both may be united
by their separatedness. Above: do you have the feeling you've
seen these men before? Do you suppose that
people who read or write books like this
appear in their nightmares?

The mind 'making': why should these quite separate images appear related? Above: this tiny bird fell from its nest in the amphitheatre at Nîmes in the south of France. A few hours later it was crushed and trampled by a crowd of people who had been watching bull-fights all afternoon. Is the sympathy we feel, for the dead bird, the bulls that die or for ourselves having read it?

Right: when the mind needs to separate itself from the world in sleep there is nothing we can do to stop it from sleeping. Similarly when the mind wishes to stay awake there is nothing apart from the use of force (i.e. drugs, etc.) that can make it sleep. But in neither case does it seem that we are able to stop the mind from 'making', i.e. dreaming

But breaking out of our habits of belief isn't easy. Let's look at the ingredients of our surroundings again. Let's examine what's happening as we sit reading this book. What do we recognize around us?

What are we aware of? Two arms? Two legs? Two ears? Two eyes? Body heat? Temperature of the air? Our weight? The atmosphere? But not its pressure. Anything else? Time passes. As we sit we may notice time passing, and that there is space between things. Things are separated by space? Space separates . . . things; or does it? Ideas and thoughts interrupt. Come back to the words on the page, read on, slow down, let the mind rest on each word, and then on what it refers to. Light, Heat, Floor, Chair, Table, Book, Feet on floor. Bottom on chair. Book in hands. Hands. Tension in muscles. Breath coming in. And out. Walls containing air, keeping in warm air, keeping out cold air. Eyes seeing this page, signals reaching the brain, hands turning the page. The mind, in its relentless momentum rushing to move on, picking up speed. Out there a book . . . in here, me. Memories of out there. Memories of this morning, yesterday. Other people—Mike, Stan. People from tomorrow, people to fear, people to use, people to enjoy. Thoughts flit in and out. The kaleidoscope speeds up. He, me, her, his, mine, thine, ours, theirs, yours, yes, no, time passing, space out there, energy locked up out there, energy locked up in here. Out there a cat sleeping. In here, ME. This something in here, thinking, wondering what to do, wondering if these words on a page are enough, criticizing that thought with another thought.

Here is the lie at work. Your mind is reading a book that I put on the page and which you are taking off it. Down here is the book, up there is you reading it. The book speaks, your mind listens. The words speak of what is normal for us when we read a book. The normal separations of the world of the senses and mind. Here we go again. Down here a book? Up there a mind? Who or what is it that knows there's a mind? A something? A knower? Maybe even a Knower? But what is really up there, mind? Time ticking by. Tick Tick Tick Tick. Seconds that tick into minutes and hours. Days divided into sets of seven. These are the ingredients of the big lie, the technicolour world of the mind that is superimposed on what is really there.

How do we lift the superimposition? How do we go beyond the limitations of the senses? How do we move?

There are two basic traditions which try to come to grips with this problem. The scientific tradition and the Eastern philosophic tradition. We will come to the Eastern tradition in a little more detail at the end of the book; but a brief note of some of its basic priorities and how it seems to relate to science may be useful here.

The Eastern tradition has taught for perhaps three thousand years that if we could obey the simple instruction—BE HERE NOW—the superimposition, the dream, would cease to exist. But regrettably, we are unable to just BE. Our minds wander; we identify ourselves with our thoughts, feelings, and sensations; we *become* them. And in doing so we cease to be truly ourselves.

As the bird moves in the air according to its fancy, without abiding anywhere, without depending on anything, so people with all their separateness move along, walk about in the mind itself like a bird moving in the air.

Opposite: Doré's engraving of Don Quixote's mental world of fear, threat, hate and horror may seem very far from 'everyday' normality—but it vividly portrays the underlying kaleidoscope of our minds' imagery

The physicists' view that reality consists of vibrations in a
fundamental field is conveyed by symbols such as this Feynman diagram
(drawn on the blackboard, right). Similarly in the Eastern tradition syllables
such as OM, pronounced AUM (inset), are held to embody the fundamental
vibrational nature of the universe

So mind needs to be stilled. We need to know it for what it is. But how?

> As a sword cannot cut itself, or a finger cannot touch its own tip, mind cannot
> see itself.

The Eastern tradition takes the view that our knowledge, including our
knowledge of the mind, is limited by the level of our awareness, by how conscious we are. This is why the approach of the Eastern tradition invariably includes in its teachings some practical way of expanding awareness.

This usually takes the form of a method of meditation. A subjective approach to objective knowledge of the mind, if you like. As the practice
proceeds and awareness grows, the illusory aspects of life fall away and there
is a deepening experience of what *Is*. In the Eastern tradition enlightenment
comes from understanding the nature of existence.

> We are constantly struggling to avoid facing what *is,* or we are trying to get
> away from it or to transform or modify what *is*. A man who is truly content is
> the man who understands what *is,* gives right significance to what *is*.
> *Krishnamurti*

It's here that the older Eastern tradition seems to be on common ground with
science, the latest and perhaps the most powerful method of entry into the
world beyond the senses. Science too claims to be concerned only with what *is*
and with giving the right significance to what is. Science one might be forgiven
for saying, is concerned to understand the existence of nature.

The simple idea around which science is built is that nature is objective, that
it can be comprehended and that the only route to this comprehension is the
systematic confrontation of logic and experience. The scientific method is un-
doubtedly a unique and powerful source of truth, a profound guide (though it
has barely begun the work of fusing the inner mental and outer material
worlds) to what *is*.

In the remainder of this book we look at a little of what the scientific
method has to show us, what it says about the nature of the material world.
First through the scientific extensions of the senses, and later through science's
combined extension of memory, thinking and mind, we follow the deduction
from limited evidence of the detail of a reality that cannot be directly imaged—the fabric of being. We look at what is revealed as the operations of
these extensions are increasingly refined, and we try to note whether, taken as
a whole, they point in any particular direction.

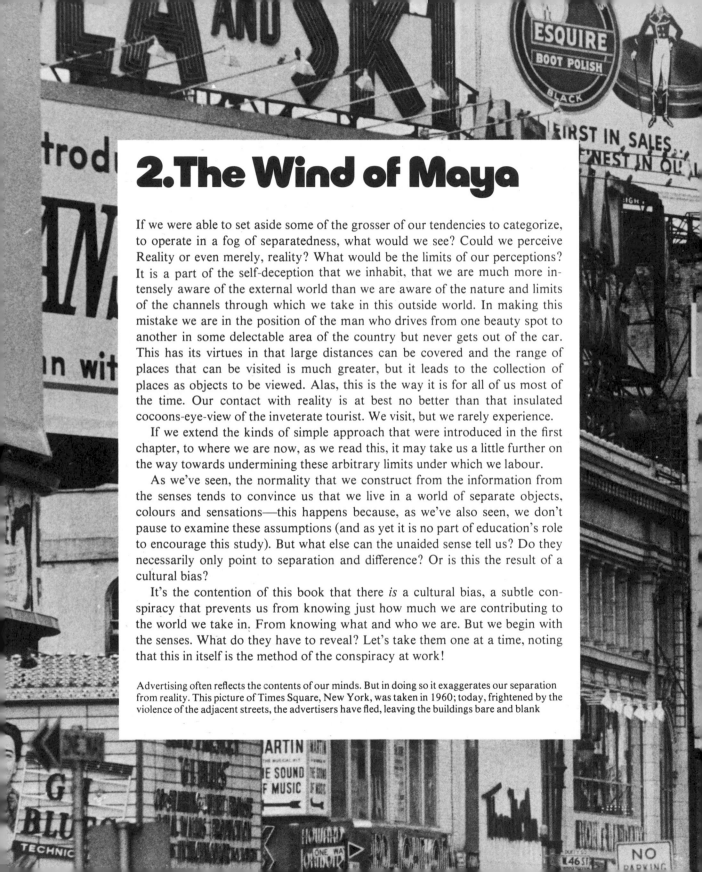

2. The Wind of Maya

If we were able to set aside some of the grosser of our tendencies to categorize, to operate in a fog of separatedness, what would we see? Could we perceive Reality or even merely, reality? What would be the limits of our perceptions? It is a part of the self-deception that we inhabit, that we are much more intensely aware of the external world than we are aware of the nature and limits of the channels through which we take in this outside world. In making this mistake we are in the position of the man who drives from one beauty spot to another in some delectable area of the country but never gets out of the car. This has its virtues in that large distances can be covered and the range of places that can be visited is much greater, but it leads to the collection of places as objects to be viewed. Alas, this is the way it is for all of us most of the time. Our contact with reality is at best no better than that insulated cocoons-eye-view of the inveterate tourist. We visit, but we rarely experience.

If we extend the kinds of simple approach that were introduced in the first chapter, to where we are now, as we read this, it may take us a little further on the way towards undermining these arbitrary limits under which we labour.

As we've seen, the normality that we construct from the information from the senses tends to convince us that we live in a world of separate objects, colours and sensations—this happens because, as we've also seen, we don't pause to examine these assumptions (and as yet it is no part of education's role to encourage this study). But what else can the unaided sense tell us? Do they necessarily only point to separation and difference? Or is this the result of a cultural bias?

It's the contention of this book that there *is* a cultural bias, a subtle conspiracy that prevents us from knowing just how much we are contributing to the world we take in. From knowing what and who we are. But we begin with the senses. What do they have to reveal? Let's take them one at a time, noting that this in itself is the method of the conspiracy at work!

Advertising often reflects the contents of our minds. But in doing so it exaggerates our separation from reality. This picture of Times Square, New York, was taken in 1960; today, frightened by the violence of the adjacent streets, the advertisers have fled, leaving the buildings bare and blank

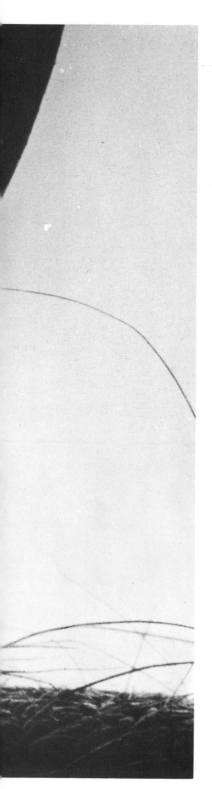

The sense of touch is always on. Or almost always. If, while asleep, we get cold, the mechanism of touch still seems to penetrate our sleeping brain. As we sit here writing or reading these lines even the notion of sitting itself can be a pointer. The body sits. The chair is sat on. There is the sitter and the sat-on. But we are held in this duet by a third element; a force holds us on the seat. We know this is somehow related to the earth, and that this action of the earth on us, which we call 'weight', can take place even at a distance. Stand up and lift your chair—now let it go—it falls. Experience tells us that wherever we do that on the surface of the earth, the chair will behave similarly. We take it for granted. But what is the mechanism that we label gravity? Kids' stuff you may think, but do we realize the extent to which, through touch, we are moulded and limited by gravity? Look at the people in the office or on the train. How many heads are off balance, how many backs bent? Gravity limits the size we can grow to be; it defines what we mean by stamina, i.e. the ability to sustain resistance to gravitational (and other) forces. If we were more aware of gravity we would be more aware of how we sit and stand and even how we lie. A major part of the consumption of our metabolic energy is dissipated in maintaining the stressful states of unbalance that 'comfortable' furniture insists are normal and good for us.

So touch introduces us to gravity and action at a distance; but touch also connects us with heat. If you are unaware of the heat of the room as you read this, it is highly likely that the temperature range of the air next to you falls between 65° and 75°F. Touch also allows us to experience action at a distance, when we warm ourselves in front of a fire or when we lie in the sun getting brown if we aren't already. But what is heat? How can something act across a distance of ninety-three million miles or even five feet to produce sunburn? Put the palms of your hands together in front of you; now rub them together vigorously twelve times or so. They feel hotter? What is it that makes them hotter? Is the heat in your hands the same as the heat that comes from the sun?

We are aware, and experience reminds us of it from time to time if we forget, that touch has limits. We cannot sustain contact with 0°C for long without first discomfort, and soon, damage. At the other end of the scale we cannot sustain contact with 100°C at all without damage. Yet a moment's thought tells us that this is only a fraction of the temperatures that can and do exist. A similar range of sensitivity exists for electricity, at the normal currents available in the car or in the home. We can barely detect 12 volts yet 100 volts is very uncomfortable and 220 volts may kill us. Yet what electricity *is* remains a mystery to most of us. It's another label, convenient as a way of describing repeatable effects, like lamps that give light when we throw the switch.

Touch is very sensitive; we are aware of, and respond to, the tip of a forefinger on a single hair on the back of our hands

Left: the sense of touch plays a crucial role in our ideas of pleasure and pain (which may often lie surprisingly close to each other in our minds). Even on a cold February day as here, a soaking with a bucket of water can be a source of great delight; but if it had not been self-inflicted, it might have been a source of considerable suffering!

Above: a London Thames-side street, flooded every four weeks by the action of the moon's gravitational attraction; everything we see in the picture is defined and shaped in some way by gravity, one of the two forces of nature we can directly experience through the sense of touch

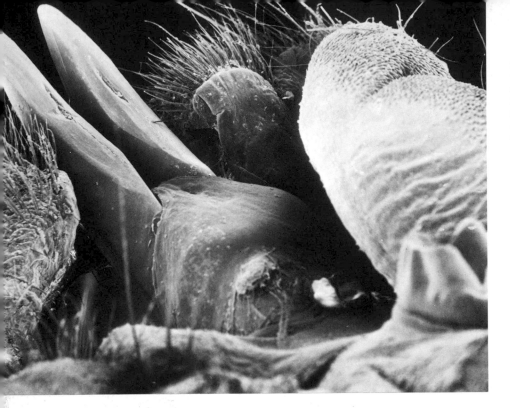

Left: this is a picture of the
mouth of a mouse. Why can't
we accept information from our
eyes just as it is? Doesn't the
mind insist on providing an
instant framework of thoughts
and opinions with which to
digest the experience?

Now in our childlike review of the sensorium, we move on to *Sight*. Suppose we have thrown the switch, or that it is still daylight and that the room is flooded with light. Here is perhaps the prime source of our urge to label. The eyes, maybe more than all the senses put together, contribute to the state of separatedness that we referred to at length in the previous chapter.

Sight tells of solidity or lack of transparency; sight originates the data which we interpret as difference between this and that colour or shape or movement and out of which we form mental models of order, chaos, equilibrium, ugliness, beauty, etc. The medium which perhaps feeds this tendency more than any other is light itself. But what is light? What is the mechanism of 'illumination'? Is it a 'mechanism'? If we have started with the idea that it will be a 'mechanism', might that stop us finding out what light really is? What if the nature of light were so subtle that we could never see 'it', but only its effects? It's our blindness to anything but these kinds of habitual thought patterns which imposes unneeded limits of mind on the transactions we enter into in the world. This is the lie at work. It tells us everything is diverse, separated, disconnected. Getting out from under the lie is getting to know what these habits of mind are, getting to know ourselves.

Of course there is a paradox here. It's the grosser forms of separatedness—pride, anger, jealousy and prejudice—that are harmful. It's not being suggested here that the more subtle function of discrimination—conscious choice—is anything other than a biological necessity. But the habit of separation runs deep. Did you know, for instance, that the eyes are not separate organs connected to the brain, but simply an outgrowth of the brain itself?

Right: the mental labels we
attach to our visual experiences
form a shifting kaleidoscope of
impressions. Here we have a
packet of cigarettes painted on
the wall of a building, but the
wall is part of one building
among several others; together
they are part of a photograph;
and whereas the street, buildings
and painting were seen by the
camera as colour we are happy
to take them in drained of their
tints, and reduced to an array of
black dots lying on the surface
of a page of a book.

Two-eyed sight insists on the perception of what we label as 'space'; we see into the 'distance', though for those of us who live in cities this is a function which is rarely exercised since the horizon may only be two blocks away at best. But what is this 'space' that we refer to in such a facile way? We feel that it has something to do with measurement perhaps, at least in the case of small spaces; but in the case of long distance we tend to think of space as having to do with how long it takes to go somewhere. Again this is more true in a city than anywhere else because mileage means less than time taken to get where you wish to go. Perhaps an equally important aspect of sight for our present enquiry is that we 'see' movement; from this mind abstracts 'change', and labels it with 'speed' or 'velocity'—or even 'acceleration', 'space' and 'velocity'. Let us note then these highly abstract labels that we have attached to the data which comes from the eyes. Each is a name for a theory based on repeatable experiences. We'll meet them again later.

The other remaining senses we may pass over more lightly. *Smell* and *taste* are highly involved with the mind function of recognition, of remembering substances, of testing for similarity or difference. But what is it in the air that we can smell but often not see? What is it that allows most people to be able to clearly distinguish the tastes of caraway and spearmint? The mind tells us that they are different, but are they really so different?

Right: a taste pore from a mouse's tongue, enlarged ×30. Though we often ignore it, the sense of taste can operate with a fine precision. For instance, most people can easily distinguish the taste of the molecule R-carvone from that of S-carvone, even though they are identical except for being mirror images of each other. The first is recogized as spearmint, the second as carraway seed (below)

Music, if we are open to it, has the capacity to flood us with delight (left) but a few tiny obstacles close to the source of the sound such as dust in a record groove can transform the experience from delight to irritation and annoyance. In a similar way obstacles close to the source of thought, ie stress, can transform our experience of life from delight to irritation.

What about *hearing,* the last of what we presently recognize as the senses? Aldous Huxley has written an account of the hearing process that deliciously fuses a multitude of differentiations.

Pongileoni's bowing and the scraping of the anonymous fiddlers had shaken the air in the great hall, had set the glass of the windows looking on to it vibrating; and this in turn had shaken the air in Lord Edwards's apartment on the further side. The shaking air rattled Lord Edward's membrana tympani; the interlocked malleus, incus and stirrup bones were set in motion so as to agitate the membrane of the oval window and raise an infinitesimal storm in the fluid of the labyrinth. The hairy endings of the auditory nerve shuddered like weeds in a rough sea; a vast number of obscure miracles were performed in the brain, and Lord Edwards ecstatically whispered 'Bach!'
Aldous Huxley *Point Counter Point*

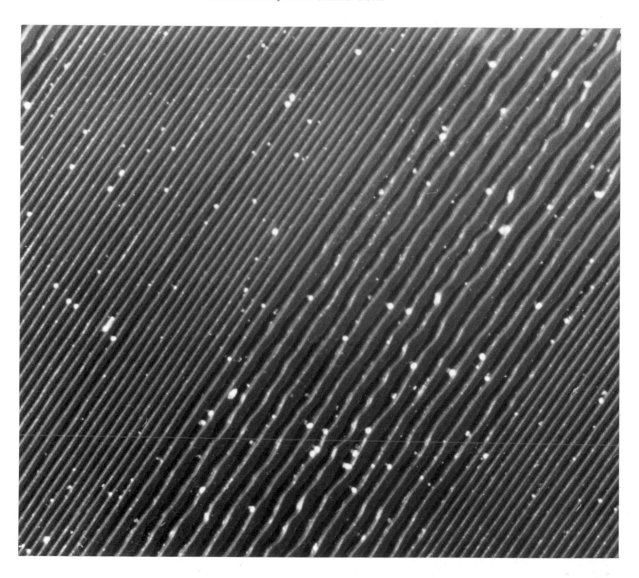

On the cosmic scale, this process of hearing (though not perhaps sound itself—there is plenty of sound *inside* solid objects that we are unable to detect) and of course all the senses, are an abnormal occurrence, as rare as intelligent life. But though as phenomena they are all rare, hearing, as an example of the defining and limiting role of the senses, can point us in a number of useful directions. First, we take in sound all-at-once, simultaneously. Secondly we may notice qualities we label 'pitch' and 'vibration' and combinations which please us which may be labelled 'resonant'. In music we notice that some sounds appear to fall on the ear more happily than others, and the mind labels them 'harmonious'. Again, a series of highly abstract labels for artifacts of the mind. We'll come across them again later in the book.

It's the business of physics to penetrate behind these labels to see if they can be reduced in number, to see if some labels can be established as facts. But we must remember that attempting to penetrate the science and physics brings its own special difficulties. It's easy to forget that science is not separated from the rest of our mental activities but is merely a highly developed extension of one corner of the kaleidoscopic turnings of mind. For most of us most of the time, scientists included, the kaleidoscope remains a kaleidoscope, a turbulent display of thoughts and feelings and sensations with the ability to form itself in the mind into paranoia or delight, self-pity or theory. Only rarely is the mind still enough for us to be able to see the husk of unconscious naming and habitual thinking that constitutes its shimmering irridescence and which so damagingly limits what we can know or be.

It's an old problem:

Hence sages who have fathomed its secret have
designated the mind as Avidya or ignorance,
by which alone the universe is moved to and fro,
like masses of clouds by the wind.
Shri Shankara acharya

As the whole world is a show, magical and unreal,
then what reliance can be placed upon it?
and what signifies pleasure or pain? Know
this egg of the world to be a phantom idea
presented for the delusion of our minds.
Sometimes it seems to be straight and at others
curved; now it is long, now short; now it is moving, now quiet again and everything in it is
continually in motion yet it seems to be still.
Ramana Maharshi

In the ocean of Infinite Bliss, the waves of
the universe are created and destroyed by the
playing of the wind of Maya.
Shri Shankara acharya

The point of this review of the mechanisms of perception has been to remind ourselves that they *are* mechanisms, and that they do colour what they transmit—and to bring into our field of awareness the fact that the world we ex-

Opposite: quietly buried among the mind's weather is memory, the storehouse of impressions, electrochemical shadows of previous experience. Why is it that some thoughts arise without effort as delight, while others are reborn under the pressure of stress as anxiety and fear?

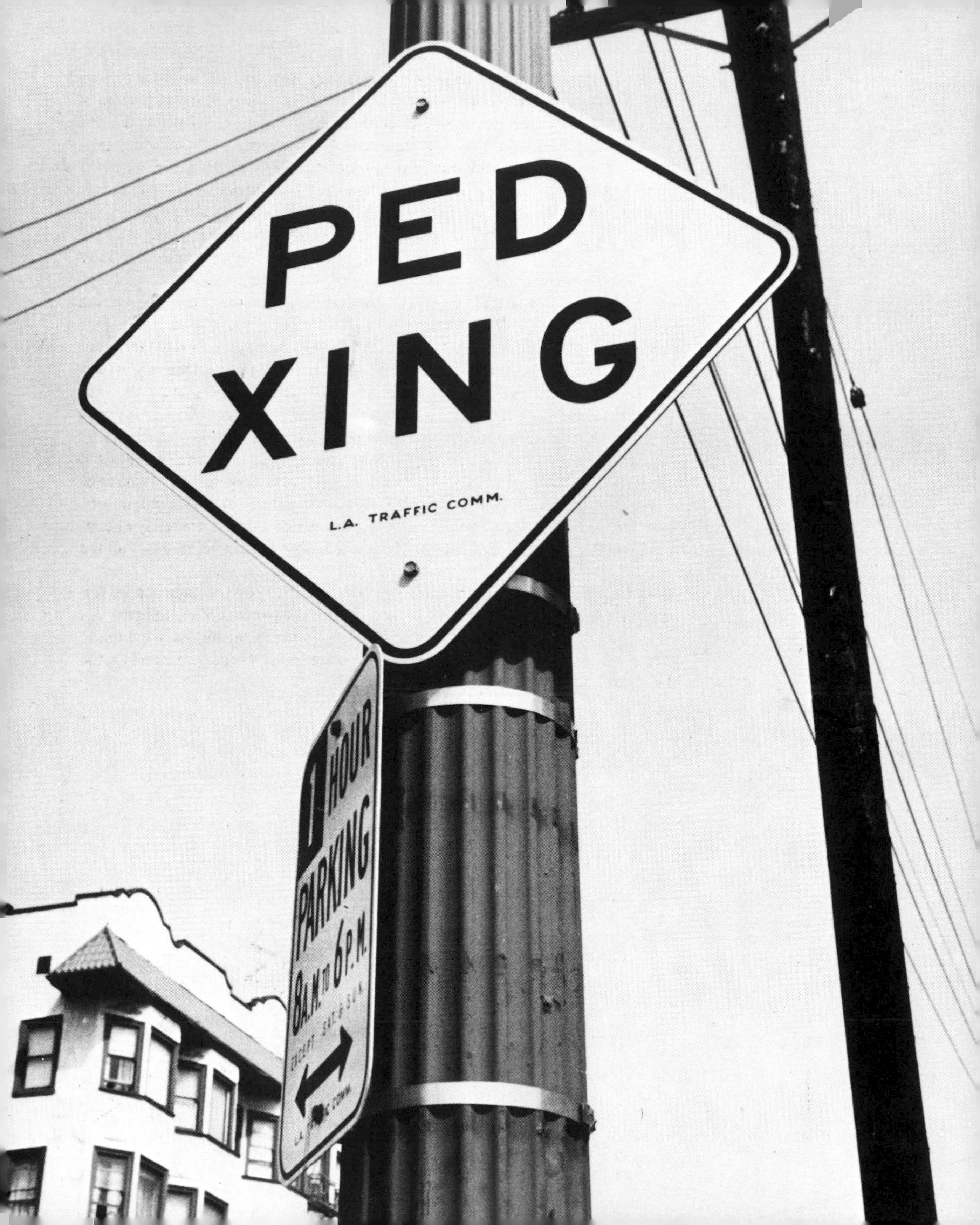
PED
XING
L.A. TRAFFIC COMM.
1 HOUR
PARKING
8 A.M. TO 6 P.M.

perience may, but does not usually, coincide with the world that *exists*. A further
reason for drawing attention to them in this way is to help de-mystify some of
the science which we are coming towards in later chapters. Science is the discipline in the West which is most concerned with establishing what does exist
and what is merely wishfulness on the part of an observer. We forget, or never
come to know, that science is no more than a range of devices—some material,
some mental—which extend one or more of the senses or functions of mind into
areas of space or time to which they would otherwise be unable to penetrate. The
images and symbols of language extend memory; computers and computer
language extend memory capacity and speed of thinking; space probes extend
the distance at which sight and touch can operate; equations extend thinking
itself.

That the world of appearances is one of images and symbols may seem obvious; a truism even. But nevertheless it shapes us. It's our own mental weather.
It has power over us. But we must avoid associating the mind only with
negativity. Nor is it being suggested that the world is not there. Our intention is
simply to remind ourselves that the mind makes a good servant but a bad master,
by drawing attention to the fact that, contrary to our private belief, our actions
are very often mechanical and automatic. This might only be one more source of
harmless comedy, if it were not for the way that memory, thinking and sensory
information, locked into habit, give us the impression that we *are* really in control. That we do direct our actions. This is the most pernicious result of all the
web of appearance.

Our trouble is that we seldom really give our attention through our senses.
We intercept, but simply do not connect with, ninety per cent of the information
available to us. Nor is it a matter of trying to improve ourselves. We already
possess all we need. What we seem to lack is the understanding, the ability, the
willingness and the strength to use it.

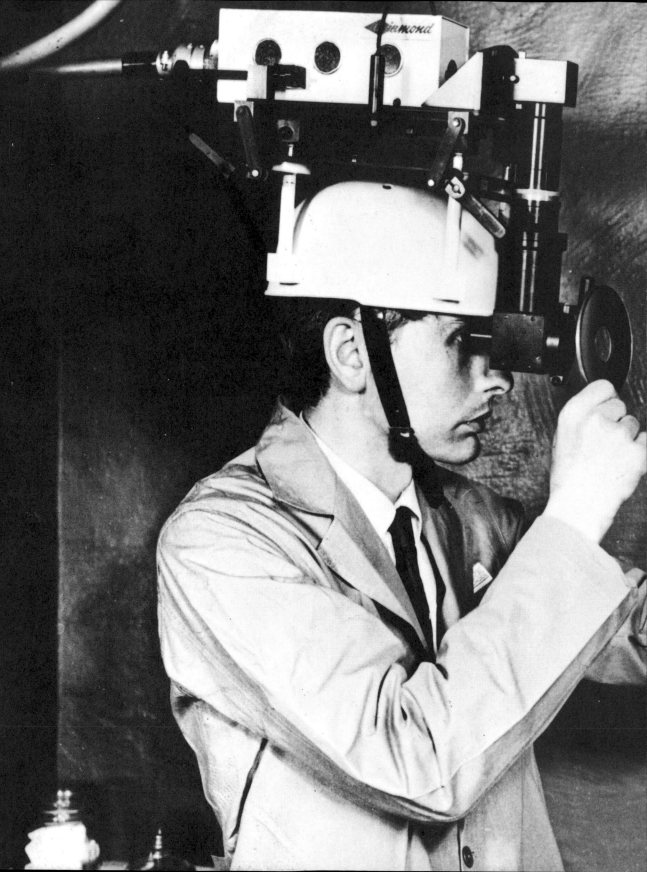

Mind, in conjunction with the senses, may be held to create a kaleidoscopic web of percepts, a piece of mental tailoring which we and the world wear and which fits where it touches, here revealing, here obscuring and teasing, much as a well conceived dress may reveal and yet conceal the female form.

But what if we go beyond the senses? If we subject the room we're in to the diligent inspections of science?

Let's imagine that we take delivery of a consignment of scientific instrumentation and equipment, a series of black boxes. We don't need to worry about how they work; many of the people who use the real things often don't know how they work.

If we take first the black box labelled 'optical microscope', and turn it on to some specimens of materials or objects from our room, what do we see?

First the diversity of the kaleidoscopic surface of the world appears extended a thousand-fold. We find a staggering multiplicity of bacterial life (and also if we look elsewhere, diatoms, amoebae, etc) and a profusion of subdivisions of larger organisms, i.e. cells. It is immediately striking that at very high magnifications everything seems to be in jiggly movement, and if we look at suspensions of tiny pollen grains in water, we see a particularly irregular movement. We might well wonder what causes this movement. If before we thought of the world as full of things, it now seems to be *made* of things. But as we look closer and longer at a wide range of subjects, some simplicity

Below: computer-generated images indicate the kinds of movement occurring within left, a liquid, and right a solid

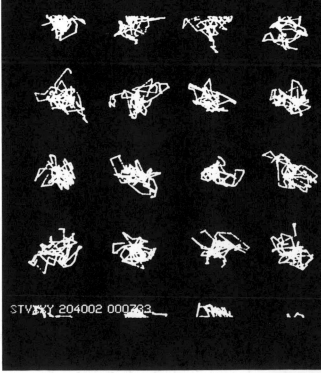

emerges from the plethora of images and shapes. We find for example that mammalian liver cells all look alike—mouse or elephant—and that the undifferentiated cells of the early stages of an embryo are similar across most of the animal kingdom. Indeed, in their basic architecture *all* living cells have greater similarities than differences.

There are also patterns of behaviour to be seen. Cells go about their business in regular ways as if obeying their own intrinsic instructions. There is a simplicity, a natural equilibrium, in the highly efficient way they lie adjacent to one another that suggests a balance between the forces operating in the cells and the space of the cell itself. And if we were to go further and combine microscopy with chemistry we might find confirmation of an even deeper, but not yet final, simplicity.

> Among the infinite diversity of singular phenomena, science can only look for invariants. . . . It was of course not difficult to see that seals are mammals closely related to carnivores living on land. It was much harder to discern the same fundamental scheme in the anatomy of tunicates (dolphins) and vertebrates . . . and it was still more a feat to perceive the affinities between chordates (vertebrates) and echinoderms. Yet it is certain, and biochemistry confirms it, that sea urchins are more closely related to us than the members of certain much more evolved groups of invertebrates, such as cephalopods, for example.
>
> *Jaques Monod*

Below: a section of a human liver cell

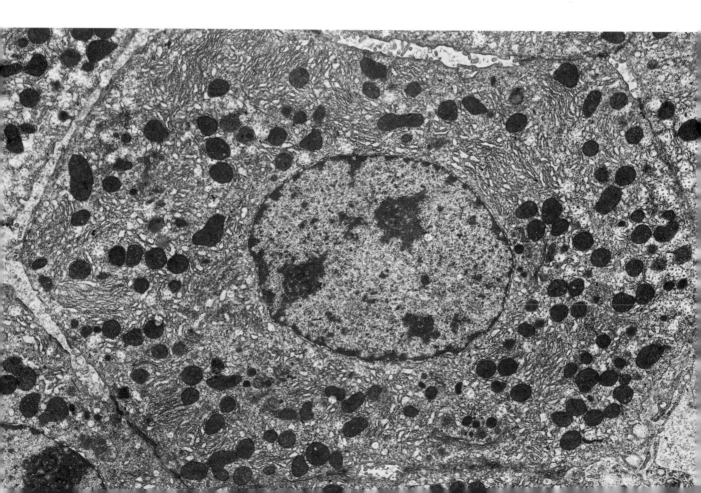

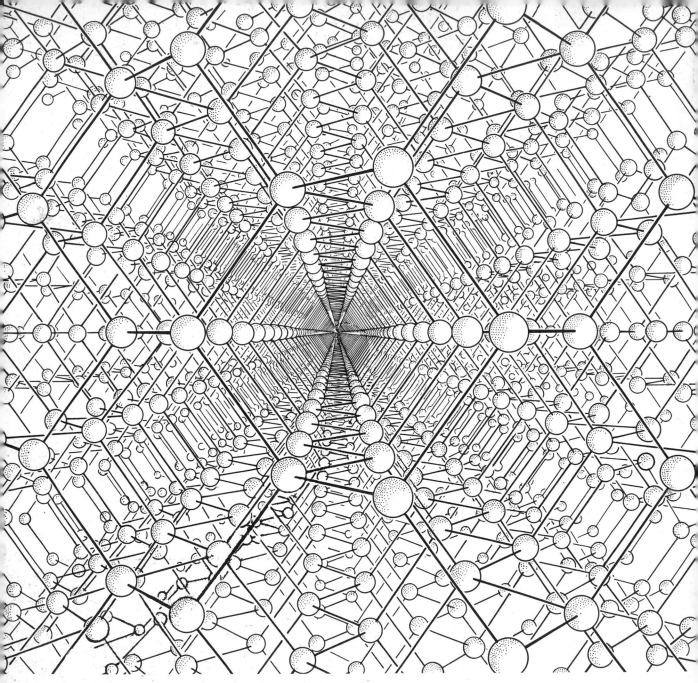

If we put some of the inorganic material of the room under the microscope, we immediately recognize many many regular forms: crystals. At first they seem to appear in unlimited variety; but we can quickly see that certain regularities are repeated. Crystals are not arbitrarily structured; certain shapes are preferred.

But as we look closer with exceptionally sophisticated types of illumination, a point is reached where the image becomes increasingly fuzzy. There is a limit to what the lens can focus. What causes this fuzziness?

Above: a schematic drawing which shows the internal structure of a diamond-like crystal. Right: crystals of citric acid and overleaf, of vitamin C, photographed using polarized light

48

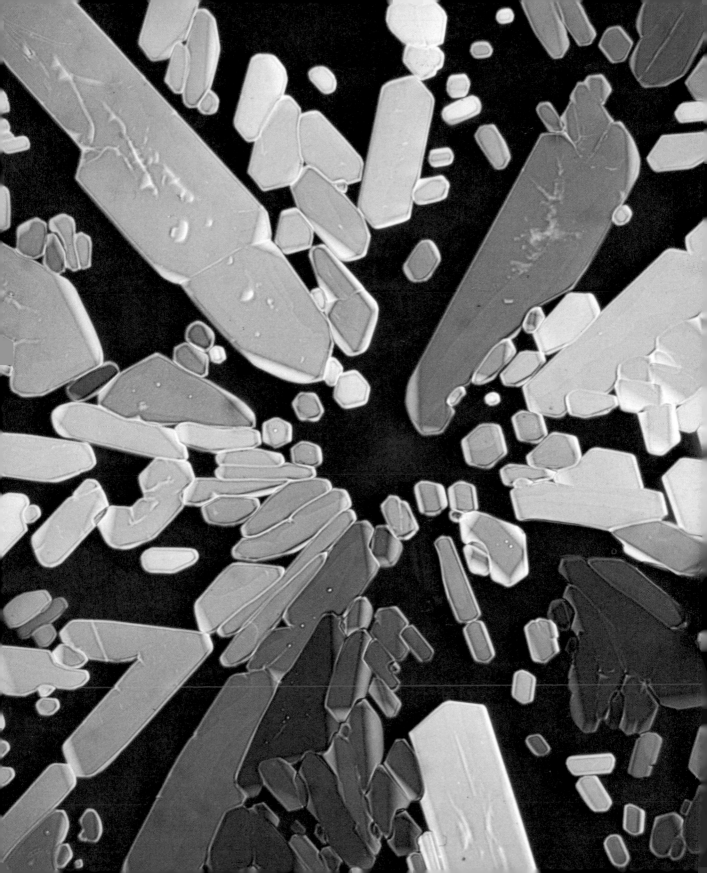

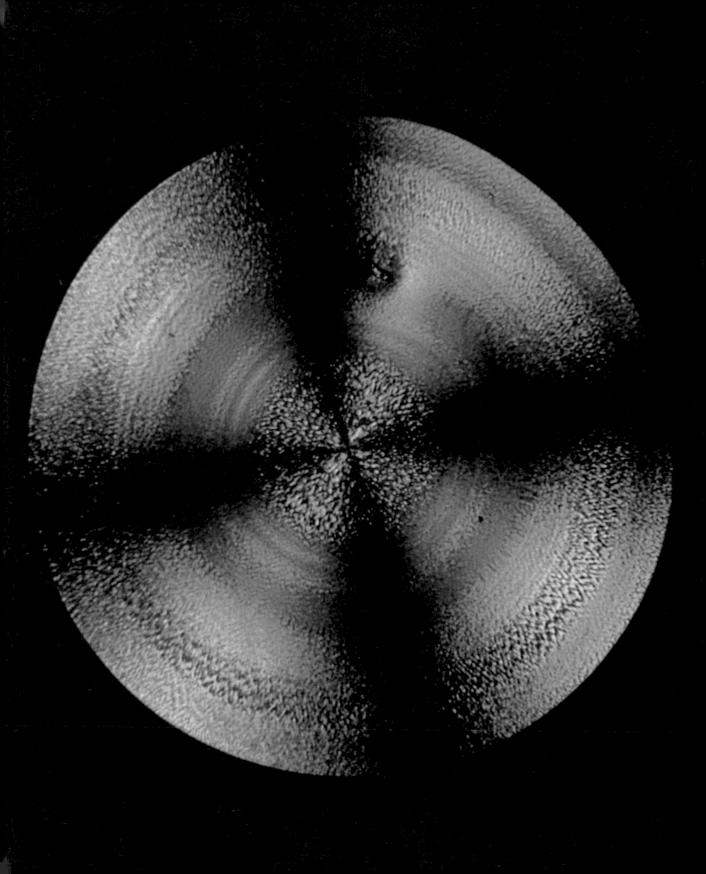

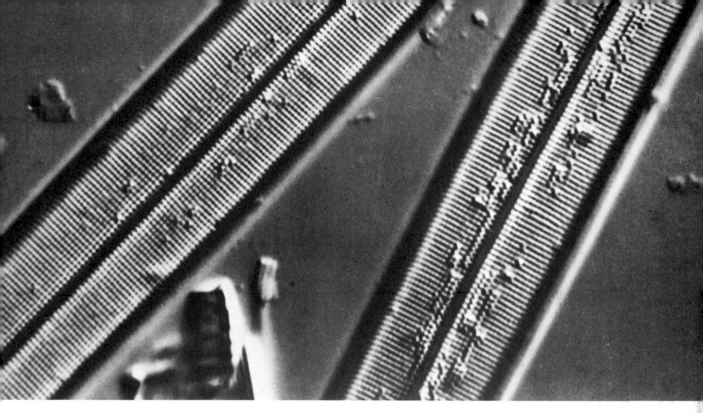

Above: the limit of resolution of optical microscopy ×2000 magnification; *amphipleura pellucida* virus photographed using Nomarski differential interference illumination. Below: a scanning electron micrograph ×25,000 magnification of *E. Coli* bacteria

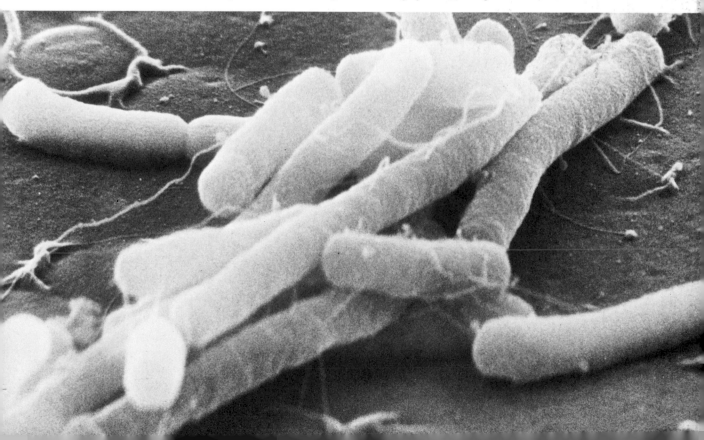

The optical microscope shows us our room as a galaxy of forms locked together in subtle patterns. Why should this be—where do the patterns come from?

If we now take our black box 'optical telescope' and turn it on the sky, a similar extension of the scale of our existence takes place. Tycho Brahe, the Danish astronomer who died in 1601, thought the stars were only about thirty times as far away as the sun. Later astronomers discovered that the nearest star is over 100,000 times more distant than the sun. As recently as 1924, Hubble conclusively demonstrated that the nebula in the constellation Andromeda does not lie in the Milky Way: this near neighbour amongst the 'island universes' is 500,000 times more distant than our nearest star. Not only

Below: the limit of resolution of the electron micrograph—a hexagonal cell of 10 carbon atoms can just be distinguished

Right: NGC 1566A, a galaxy in northern Dorado, taken by the ESO Schmidt telescope

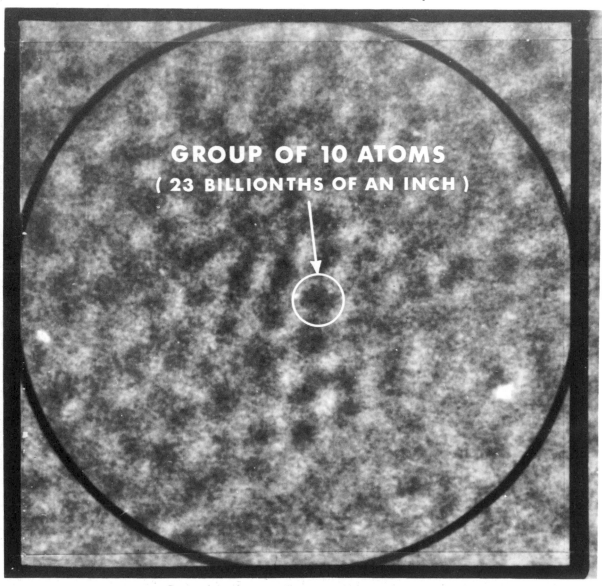

GROUP OF 10 ATOMS
(23 BILLIONTHS OF AN INCH)

is the distance extended, but the numberless stars of our galaxy turn out to be themselves only a part of one galaxy among numberless other galaxies.

Yet our black box also leads us to see that within the unimaginably vast quantities of substance visible through the telescope, there is order. The galaxies have pattern, and so also do the planets of the solar system. A force which we have come to call gravity defines planetary orbits so exactly that we are able to send Mariner Nine on a journey of 149 million miles to Mars and get close enough to be able to send back images of its surface. And in the galaxies themselves, gravity lends form to space, as stars condense within them and maybe in turn spin off masses of planetary material. And yet there seems to be a still more simple shape to the relationships between the galaxies. In addition to the connections defined by gravity, it appears that the galaxies are moving away from each other, apparently the fragments of a single gigantic explosion. Yet is it imaginable that there could once have been a moment with no other moment preceding it?

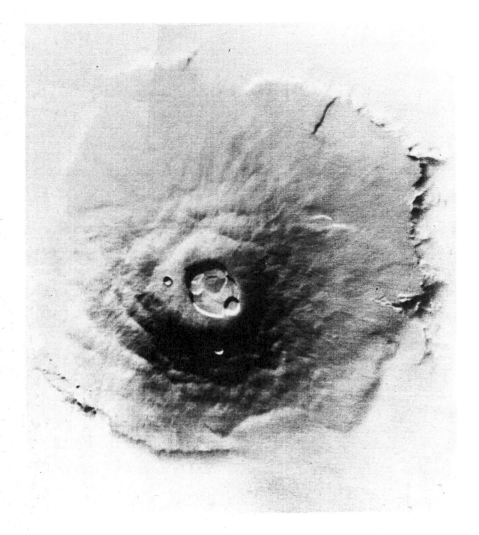

Left: Mariner 9 picture of Olympus Mons, a volcanic mountain on Mars 14 miles high. Mariner 9 was 149 million miles away from Earth when the picture was made

Right: Mariner 10 picture of a ring basin on Mercury, 800 miles in diameter

Pictures of a star field near the South Galactic Pole, taken by ESO telescope, show present limits of earth-bound optical telescopes. Indistinct spots on the ×100 enlargement (below) are distant galaxies

Answer that convincingly and it's instant Nobel prize for you. For the moment it is enough to recognize gravity as the constant force by which the earth 'knows' the moon is there and vice versa, and which defines the allowable relations between the stars and planets, and which as well as defining the pattern of what there can be, determines the evolution that there can be of that pattern.

But just as we found that our optical microscope black box has a limit beyond which the image is fuzzy, so do we find with our optical telescope. Why should this occur? Of course you probably know or have heard that it's because light itself has structure; that it occurs in the form of waves. Light is energy patterned as waves, vibrations in space, or if you like vibrations *of* space.

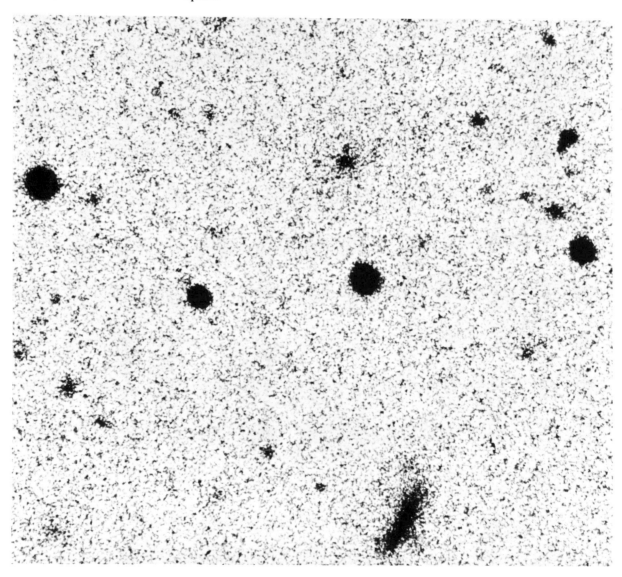

Light waves have special qualities in that they always travel through space at exactly the same speed. Furthermore, for reasons that we will come to later, the individual waves behave as though they were 'things', tiny pieces of light that have been called photons. The reason that optical telescopes and microscopes have natural limits of resolution is that light waves are only about a 2000th of a millimetre apart, so that optical images are always spread out or diffuse by at least that much. And magnification can only make things worse! (It's rather like trying to draw a picture with fine detail using a blunt pencil.) Is there anything that can be done about this? The answer with respect to visible light is no. Different colours of light do mean different wavelengths—red light is longer in wavelength than blue, but they are both close to 2000th of a millimetre in width.

The way out lies in the fact that visible light is only a part of a much larger spectrum of *invisible* 'light' waves. For instance if you think of visible light as one octave of eight notes on a piano (frequency doubles with each octave), then it is only one of at least thirty octaves of invisible 'light' for which detectors exist, some very much shorter in wavelength than visible light, and some very much longer.

But what happens if in examining our room we use light which is of a wavelength either much shorter than that of visible light or much longer? If we first look at the 'light' that is only a little longer in wavelength than visible light, using our 'thermograph black box', we find that images of the room can be made which are *heat portraits,* pictures of an aspect of the three-dimensional field of temperature of the room. Our eyes are adapted in a highly special way to only a limited range of the light spectrum. To a snake such as the pit viper, which sees in the dark by sensing infra-red, our room would perhaps look like a swirling fog of temperature difference in which the people and hot objects glow with radiated infra-red. We may not like to recognize it but, if we could only 'see' it, our constant state resembles more exactly that of the race-horse steaming with exertion after a race! Similarly, if we used our thermograph black box to look at our cities from near space we would see them at night as glowing with heat. In other heat images the infra-red heat reflected back from the earth's surface is detected and combined with a certain amount of reflected visible light, to give a wholly new kind of portrait of our surroundings.

Right: the electro-magnetic spectrum. Illustrated here are 30 octaves of electro-magnetic radiation-frequency, (wavelength doubles with each octave). Visible light occupies only a part of one octave. To represent the whole of the spectrum that has been detected the piano would need to be drawn another 20 octaves wider

Overleaf: a thermograph or heat portrait of a lady smoking a cigarette. Heat (infra-red radiation) is *emitted* by her body—red is hot, blue is cool

Overleaf top right: E.R.T.S. false colour photo of San José, California, a composite print utilizing *reflected* infra-red radiation. Vegetation, which strongly reflects infra-red, is recorded as red

Overleaf bottom right: the author's children and friend: a snapshot on an infra-red false colour emulsion with visible light filtered out: this also records reflected infra-red from vegetation as red

gamma rays
nuclear radiation
X-Rays

ultra violet
visible light

infra red

T V

FM radio

AM radio

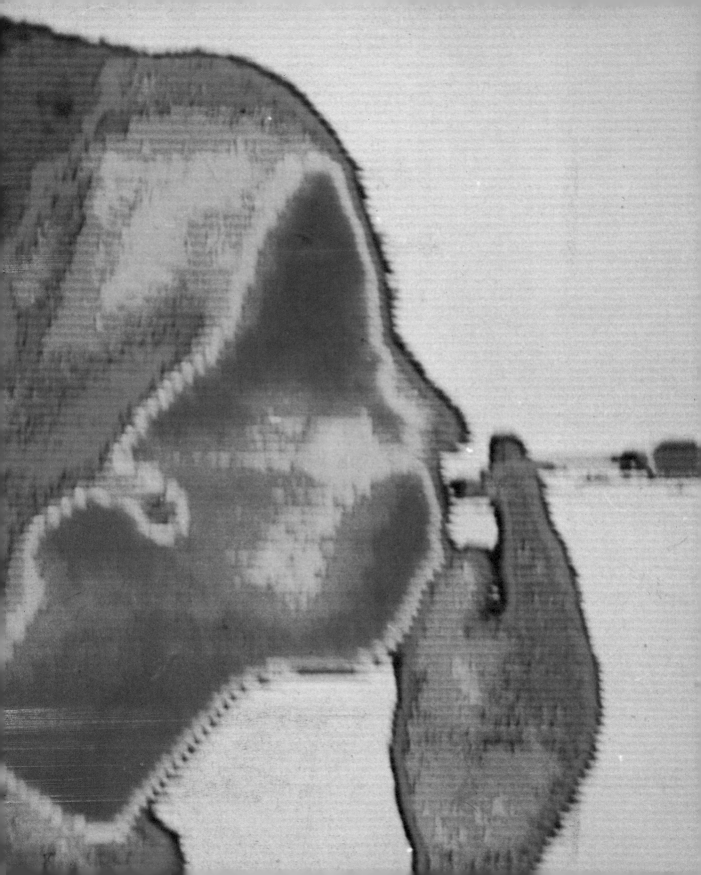

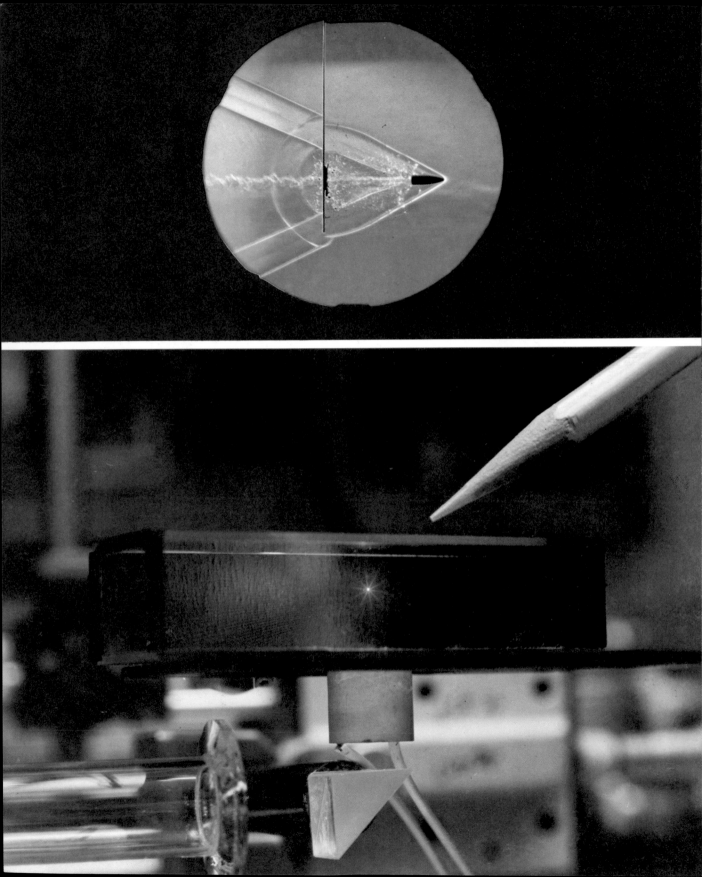

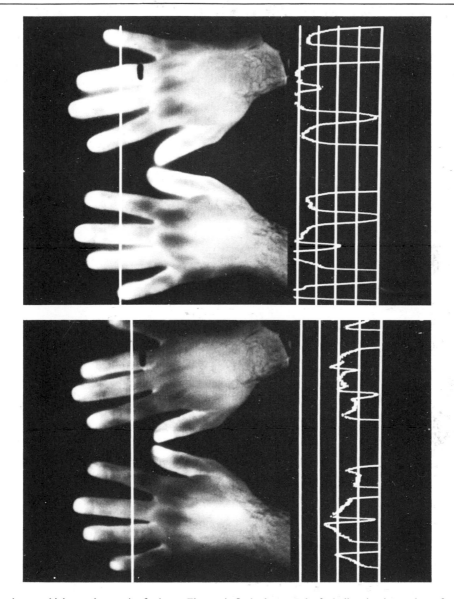

Opposite above: a high speed portrait of a bang. Electronic flash photograph of a bullet piercing a piece of card. Note shock wave beginning to be reflected from the target

Opposite below: a glass sphere 1/1000th of an inch in diameter held aloft by a beam of light from a laser. The laser beam, not visible in the photograph, passes through the lens at the end of the glass tube and is bent by the triangular prism to focus on the glass particle. The particle rides the intense beam of coherent (all the waves in step) laser light much as a table tennis ball can ride a fountain jet of water

Above: high-resolution thermograph shows the effects on hand temperature when a non-smoker smoked and inhaled one cigarette. The top picture was taken before, and the lower one 10 minutes after smoking. The white line temperature scan through the hands reads out temperature on a temperature profile scale at right. Temperature range is 30–34°C. The change in temperature is presumed to result from the constriction of blood vessels caused by nicotine in the tobacco

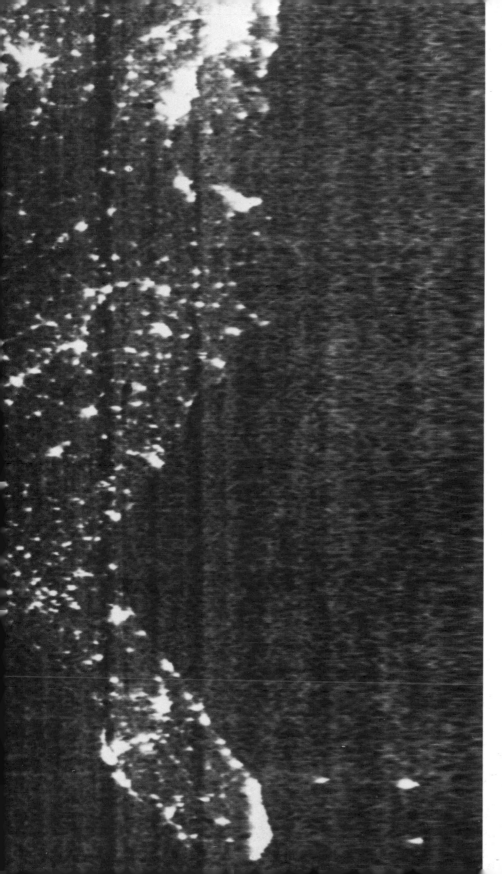

A U.S.A.F. meteorological satellite photo showing the Eastern half of the United States during the night of the 13th February, 1974. Florida is towards the bottom right. The bright spots are the heat being emitted by the centres of population

If we care to pause here we'll begin to realize just how specialized our sight is. It generates very precise information, but within only a very limited range.

Now, if we switch on our next black box, for detecting even longer wavelength radiation (a radio receiver), we find that every corner of our room is permeated by this kind of radio-frequency 'light' and our special receiver will also be able to pick up radio waves from the dial tone, which are radiated from the telephone wires in the room, and similarly, a more powerful signal that radiates from the mains electric wiring when a circuit is in use. If we were to use the black box field-strength meter we would be able to see that these fields of radiation pulsating in the room contain energy. By tuning our receiver, i.e. limiting it to only one wavelength, we would find that some of the radio waves had perhaps originated in Peking, but were still strong enough to penetrate brick and stone. What is it that travels so far and yet remains so strong? Is it a thing? Or a force? Or what?

Our collection of black boxes also includes a radio telescope (in a special microminiaturized version not yet on the market and never likely to be). If we turn this on the sky, then we discover that not only do the sun and stars put out huge amounts of radio waves, but that the sky is thickly populated with objects that are only visible on earth through their radio 'light' output. By reckoning how far these radio waves must have come, and therefore how much energy they must have lost, it has been calculated that in terms of energy some of these radio stars make the sun look like a candle.

As well as these radio objects, there is a huge amount of undifferentiated

Below: a side-looking radar photograph of an earthquake area—the San Andreas fault in the hills south of San Francisco. The long straight line at the right is the Stanford linear accelerator

Right: a section of a chart of radio sources made by the University of Cambridge one-mile telescope. Each set of peaks is a radio galaxy or quasar

'noise' (a mixture of many frequencies long and short) coming from space itself. All this wave activity is impinging on us, apparently undetected by the normal senses.

Now let us swing right over to the other end of the spectrum and put into use the waves that are much shorter than visible light. We might expect that using smaller waves such as these we could just simply make 'invisible light' microscopes. In fact this is not feasible, because as the wavelength of the light gets shorter the energy of each photon increases. At wavelengths not much shorter than visible light the energy of the photons is such that *each* photon is so energetic that only a few are enough to break apart vital molecules and cause cell damage in tissue specimens and deterioration in other materials.

The first invisible 'light', shorter in wavelength than visible light is ultra-violet, with which we are fairly familiar. It's the agent in the sun's light that initiates the process leading to sunburn, and is also the principal active ingredient, along with oxidization, in the ageing process of paints and plastic surfaces.

A more revealing black box is the one which manipulates the even more energetic shorter wavelength radiation we know as X-rays. X-rays don't occur naturally in our room (though they are plentiful in space they don't penetrate the protective atmospheric cocoon). But they can penetrate our bodies. If we use our black box to produce a very fine beam of X-rays, and if we direct it at a very regular structure such as crystal, we find that the single fine beam is broken up into a set of spots as if the crystal consisted of many small points rather than a solid piece of material. What is it that causes this array? It's not a direct picture of anything, but as we will come to see again and again as we go on, it's an indirect image of the regularity of structure of the material being looked at.

There are no prizes for guessing that the patterns are caused by the X-rays being scattered from the basic building blocks of the structure of the crystal: *atoms*. Other less simple groupings of atoms, molecules, also scatter X-rays in this way, and from the 'diffraction patterns' they make, very detailed pictures of the structure of molecules have been deduced.

Another of our black boxes, the Field Ion Microscope, can give us a direct image of the arrangement of atoms of some substances. It uses beams of atoms to make images of groups of atoms in crystals. Unfortunately the huge amount of energy it pours into the act of looking very quickly evaporates the material being looked at. For this reason, which is a problem we will meet again, only very high melting point metals can presently be imaged. But in them we can see directly the fineness of structure and the baroque richness of texture that pervade all crystalline material.

We are, the world is, an atomic structure. Through their affinities atoms bond sometimes strongly, sometimes weakly, sometimes not at all. All the diversity of substance of our room discussed on the previous pages (and the pages themselves), can be accounted for by the properties of atoms. But what is an atom? We've all heard the word and we may be filled with confidence that we know about it, or know enough about it. But unless we are very lucky our mental image of it is likely to be somewhat archaic. Because in our study of the world behind appearances we have reached a definite threshold with the

Right: X-ray photograph of the author's lungs circa 1960

Below: X-ray back-reflection Laue photograph of a single crystal of silicon. This 45-minute exposure is not of course a direct image of the crystal, but of X-rays diffracted from it. The crystal's structure can be inferred from the array of dark spots

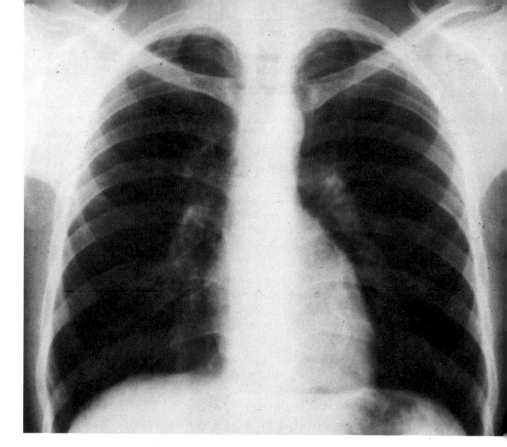

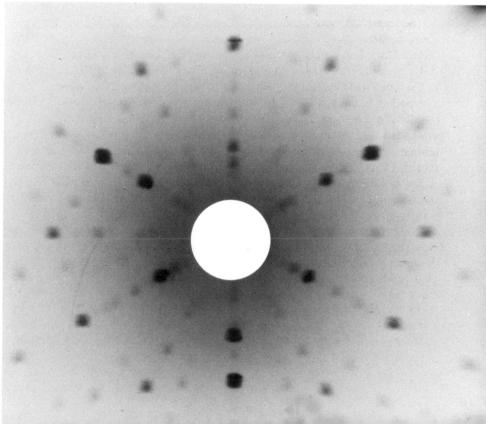

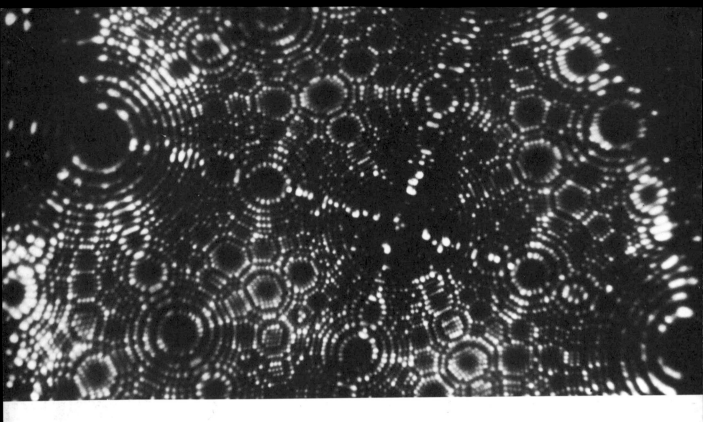

Field Ion microscopy is presently limited to revealing the structural patterns of metal crystals. Each of the white dots corresponds to one atom in the metal crystal. Above: a crystal of pure gold. Below: an aluminium crystal. Opposite: an excellent model of the way atoms stick together can be made by creating a large number of identical bubbles on the surface of a liquid (see p. 73). The surface tension forces between the bubbles that cause them to form regular arrays are analogous to the forces between atoms in metal crystals

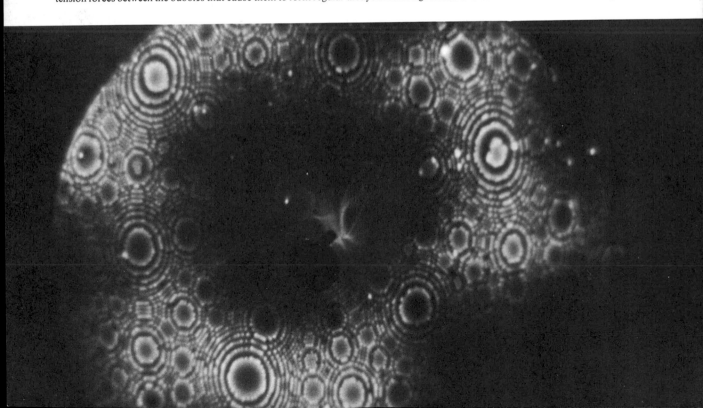

Opposite above: an oblique view of a raft of identical bubbles on the surface of a soap solution, showing the re-crystallization of the bubbles after annealing (i.e. stirring)

Opposite below: a photograph × 200 of the surface of a piece of brass, showing the similar recrystallization after heating to 1050° for one hour. The metal was etched in ammonium hydroxide and hydrogen peroxide to reveal the re-crystallized grains

Below: bubble raft model. The pictures of bubbles opposite and on the previous page were taken using a small tank, pictured here, based on the original model suggested by Sir Lawrence Bragg, FRS.

Rafts can be formed on liquids such as detergents, but the bubbles tend to break down rapidly. The solution recommended for the formation of the bubbles is 15·2cc of oleic acid (pure redistilled) well shaken in 50cc of distilled water—mixed thoroughly with 73cc of 10% solution of triethanolamine, and the mixture made up to 200cc. To this is added 164cc of pure glycerine. The whole mixture is left to stand and the clear liquid is drawn off from below. (Sir Lawrence Bragg's paper describing the model is reprinted in *The Feynman Lectures on Physics* Vol II, Addison-Wesley, Chapter 30, pp 10–26

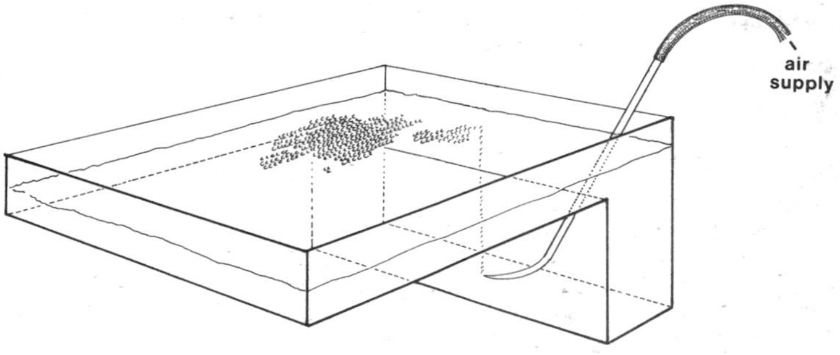

air supply

Think of each white dot in the Field Ion picture as an atom 10^{-8}cm in diameter (there are about 250 million atoms to the inch). Now think of each atom as being the size of London's Albert Hall (overleaf). Think of the space as being filled with resonating electron waves. On this scale the nucleus down in the centre, which contains 99·999% of the total weight of the atom, would only be the size of a grain of salt

atom. We are now treading dangerous territory littered with man traps. It's here that we step across the boundary between the directly perceivable 'real' world and into the shadow realm of deduction and mental modelling.

To get a taste of what the most up-to-date model of the atom means, we have to start to use analogies. The one we'll use for the atom is not very complete nor does it pretend to be exact (we will meet the atom again in the next two chapters). For the moment we need to update our image of the atom, if we have one.

The atom is basically very simple. Look at one of the field ion micrographs of gold again; look at just one of the white dots. Now ideally, holding that white dot in mind, you should go to a big hall like the Albert Hall in London or the Houston Astrodome or the Pallazzo Del Sport in Rome (see the picture on pp 74–5). Picture yourself sitting in the middle of the space. First think of the whole volume of the place as an atom, one of the dots on the field ion micrographs. Now, think of the whole space of the hall as filled with 'electron-ness'. This electron-ness is the manifestation in the atom of those effects that physics has labelled 'electron'. The electron may be thought of as an electrical condition of space. If we were able to isolate the 'electron' it would be a thing, but it's more exact to think of it within the atom as a cloud of probabilities. Don't be frightened of this. It may seem vague but as we'll see later there is a precision in the vagueness. And remember your body sitting there is a cloud of electron-ness too.

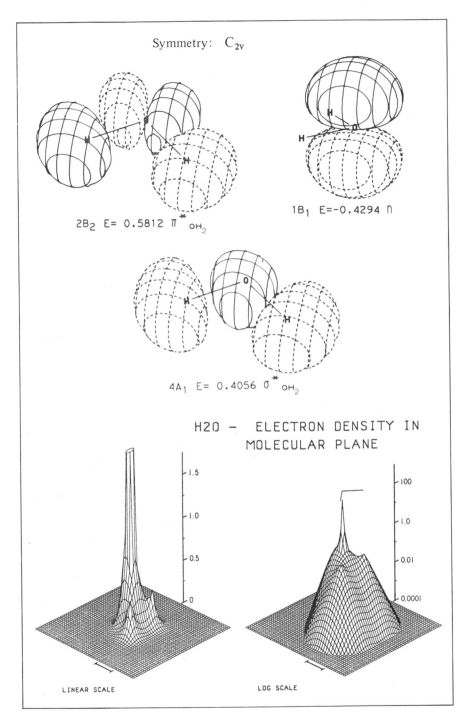

Symmetry: C_{2v}

$2B_2$ $E= 0.5812$ $\pi^*_{OH_2}$

$1B_1$ $E=-0.4294$ Π

$4A_1$ $E= 0.4056$ $\sigma^*_{OH_2}$

H2O — ELECTRON DENSITY IN MOLECULAR PLANE

LINEAR SCALE

LOG SCALE

With the atom we pass beyond the limits of what can be seen. The detail of the water molecules in mist (above) cannot be directly imaged. Direct perception of them must be replaced by inference and the mental model. But precise partial models can be made; opposite (above) are computer-drawn images of the shapes of water molecules of varying energy levels. And below them are computer drawn images of the *densities* of the electron waves in a water molecule

Depending on what kind of atom our concert hall model is, there might be from one to ninety-two separate electrons in the cloud. Each of them a ripple in the electron-ness that fills the hall. These electron ripples are being held,

throbbing, in a resonant three-dimensional shape by the forces exerted through a tiny nucleus at the centre of the atom. Remember the size of the concert hall? If it was a hundred metres wide, the nucleus on that scale would be about as big as a grain of salt, or the full stop at the end of this sentence. And yet the nucleus contains 99·999 per cent of the total weight of the atom.

What is happening in the atom is that there is an electrical interaction going on between the nucleus, which is always electrically positive, and the electrons or electron-ness which are electrically negative. It's this interaction that structures the electrons into a state of equilibrium in which they become regular, strong electric fences. The strength of the fences is what you feel if you tap your finger on the edges of the chair or table. It's this delicate balance of opposing forces which stops you falling through the floor.

The forces that hold atoms together are partly attractive, partly repulsive. At longer distances the forces are weakly attractive; at short distances they are quite strongly repulsive. It's as if the atoms in a molecule were held together by two kinds of spring: one weakly in tension, one strongly in compression. In this sense a large molecule somewhat resembles a mile-high interior sprung mattress. What happens when something breaks is that the atoms are pulled apart from one another until the weakly attractive forces, which get still weaker with increasing separation, just aren't strong enough to hold the two bits together (against gravity, or the wind, or whatever forces may be about). In fact if two carefully machined flat clean surfaces are held in contact they will, in favourable circumstances (when the inevitable ridges and furrows on their surfaces are ground away), fuse together.

Though there are only about ninety-odd different kinds of atoms, there are thousands of different ways in which they can bond together. Just these bonding possibilities can account for the whole of the diversity of texture, elasticity, viscosity, hardness, adhesion, chemistry, and so forth of all the materials of the entire world.

So we have come a little way. We can see that the whole of the world that we take in with the senses, and also that beyond the senses, is structured as objects because of the nature of the forces in or between them. Thus forms, arrays or patterns of only certain definite kinds can arise. We might well conclude from this that the world is just as we perceive it, only more so. With the eyes we see many 'things'; with a variety of devices to extend the senses we see at first more things; but then we begin to see a subtler level of simplicity.

Should we rest here? Do we know enough? The properties of the atom do account for almost all the phenomena of the day-to-day world. Especially shape and structure. Perhaps we could stop here, but many questions remain. Why does metal get red hot, not green hot? Why is the sky blue, not green? What keeps the sun going? To be able to answer these questions we have to give up one of our most cherished possessions, the idea that we are independent beings, 'things' in a world of other 'things', whether the things are minds, bodies or atoms.

We have to step up to a more subtle level, to a reality that cannot be 'captured'—i.e. separated by the camera—but which must be deduced in its wholeness, by the mind.

So now we have some sense of our room as a structure of forces, of our bodies as conglomerations of cells built of molecules, built in turn of atoms built in turn of—what? 'Particles' is the conventional answer. But there are dangers in this simple statement of fact, since it is part of our cultural heritage to assume all kinds of things about particles which are not in fact the case. The next chapter will deal with this problem in more detail.

What is a particle? Here's a way of coming to it. Try the following little mental experiment. Close your eyes, and let your mind rest on just space, an idea of just space, three-dimensional emptiness. All right? Now think of the space as locally distorted, unevenly distributed(!), pinched up, concentrated into pointlike ripples of energy only 10^{-13} cm in diameter. The intensity of this concentration of energy *is* mass; it has *weight*.

Here's another less exact way to come at it. Think this time of a thin sheet of rubber (a piece of two-dimensional space) held perhaps in a frame, but not stretched tight. Pinch between your fingers as small a piece of rubber as you can and twist it round. Now think of the twist as a particle. Can you see that the energy it takes to hold the twist in place in a way *is* the particle? And also we should keep in mind that each particle, each twist, is a property of the elasticity of the whole piece of rubber.

It seems contrary to common sense doesn't it? These pinched-up ripples in space have another non-sense quality, they behave simultaneously as objects and also as waves. Did you get that? Particles are simultaneously 'waves' and 'things'. It depends on what we use to look for them: a 'thing' detector finds objects; a 'waves' detector detects them as 'waves'. But they are the same.

They have another quality we might as well come towards right now and that is that they are also simultaneously mass(ive) and energy. In other words, if we use an energy detector they register as energy; while if we use a weight detector they register as mass. Again there is no contradiction; all energy has some mass and particles are highly concentrated pointlike ripples of energy. And the concentration is so intense that they can have weight.

Now where does this find us in our study of the room we're in? We've found that the room is bathed in light, some of it visible, most of it of radiation which we do not seem to sense directly. But this radiation is itself energy, in that it is also wavelike particles, photons; concentrated energy that in this case is structured by the fixed velocity of 186,000 miles per second at which the photons are propagated.

Apart from this ebb and flow of energetic radiation, we may absorb the idea that we *are* energy, that our structure is formed of energy itself acting in the fields of force across space. The only difference between this book and that body of yours is how the energy is organized. In a very ancient Indian text, the Chanda yoga Upanished points out the significance of this knowledge. 'By knowing a lump of clay you know all things made of clay; they differ from one another as it were in language and in name, having no reality but their clay!' Do you see what this means? It's a fact to be realized internally. Personally. Take it to heart. This *is* that.

If we can begin to realize what this may mean, it will take us towards the deeper subtleties of the physical material that we are and in which we move. In

The drawing opposite suggests the tension induced in a two dimensional surface by twisting (see text)

Previous pages: the drawing symbolizes the paradox of a particle that is also simultaneously a wave

Man has no body distinct from his Soul;
    for that called Body is a portion
      of soul
    discerned by the five Senses,
      the chief
    inlets of Soul in this age.

Energy is the only life and is
    from the Body;
    and Reason is the bound or outward
    circumference of Energy.

Energy is Eternal Delight.

William Blake
The Marriage of Heaven and Hell

time, we may perhaps be able to come to know it and ourselves as a kind of soup of particles, in which the differences are of flavour rather than kind. The soup is a thin one – the average density of matter in the visible universe is only about one atom per cubic metre – though for some reason the average density of energy in our own galaxy is about a million times higher than that of the universe as a whole. Of course this cosmic soup is not static, but a dynamical process in which everything is in movement. On the cosmic scale, gravitational energy is the predominant form, far surpassing heat, light and nuclear energy. Most of the energy flow of the universe as a whole occurs as the gravitational contraction of very massive stars, converting vast amounts of energy into light, heat and the energy of motion.

Perhaps from this we can begin to sense the surges and the ebb and flow of the cosmic forces as they enter into and leave these perturbations of the fundamental fields that are *us*. What we call life is just this ebb and flow of excitations, as energy enters and leaves us, flickering in and out of quivering atoms like a flash of a peacock's tail. This excitation, this interchange and exchange of energy, is the fundamental behaviour in our room.

Just as the atomic structure is an equilibrium of all the possible states which the forces pervading the structure have the possibility of being, so the exchanges of energy within and between the atoms themselves are subject to an accounting system of rules of what is allowed and what is not allowed, and everything that can happen does happen. Thus the *physical structure* that has been deduced from observations also has a *structure of behaviour* within which it operates. Eddington wrote about it years ago.

> The external world of physics has become a world of shadows. In removing our illusion we have removed the substance, for indeed we have seen that substance is one of the greatest of our illusions . . . The sparsely spread nuclei of electric forces become a tangible solid; their restless agitation becomes the warmth of summer; the octave of aetheral vibrations becomes a gorgeous rainbow.

But is there anything within this shimmering vibration of total transience that is constant? (Constant meaning literally always and forever.) Atoms are not constant. If enough energy is used they can be split apart into electrons and nuclei with the production of (transmutation of) large amounts of energy as photons. As for the electron, there is a high probability that it will find another nucleus to lock into; perhaps one which is not altogether bare, but merely ionized, short of an electron to make up its electrical balance. What is constant about the electron is that it's a particle which carries a unit of 'electric charge'. (It should be added that physics does not know what 'charge' is. It is a label for a property which remains constant, i.e. is subject to observable rules.)

Let's look at the nucleus. Is it constant? What do we find? From one to over 200 nucleons locked tightly together. All whirling round at speeds of perhaps 100,000 kilometres per second in a space only 10^{-13} cm in diameter. But the nucleus too can be broken apart. It turns out to consist of only two kinds of particles, protons and neutrons. But the nucleus can be broken only at the cost of a great deal more energy than it took to break apart a molecule. A hundred million times more energy. The extra force is required because the

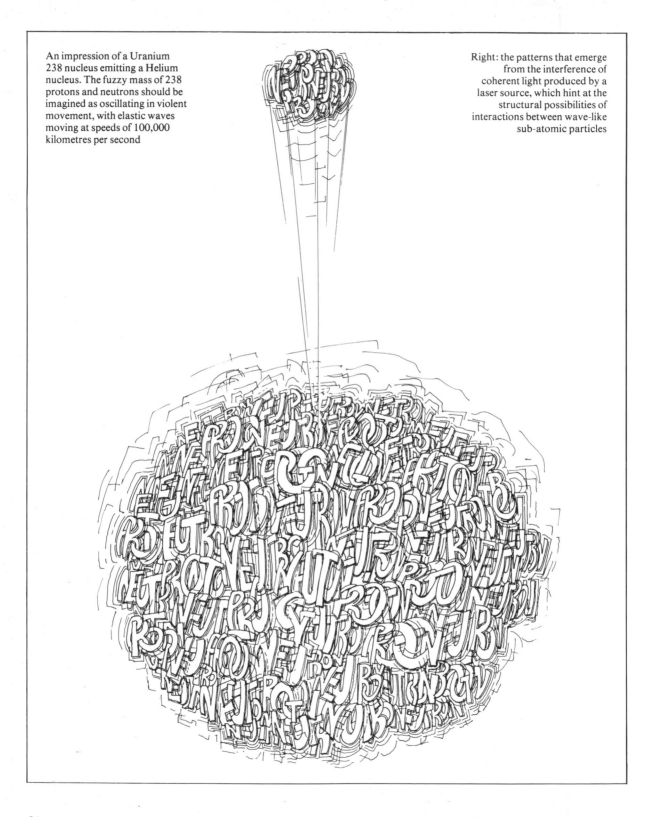

An impression of a Uranium 238 nucleus emitting a Helium nucleus. The fuzzy mass of 238 protons and neutrons should be imagined as oscillating in violent movement, with elastic waves moving at speeds of 100,000 kilometres per second

Right: the patterns that emerge from the interference of coherent light produced by a laser source, which hint at the structural possibilities of interactions between wave-like sub-atomic particles

protons and neutrons in the nucleus are glued together by a tremendously strong force. In fact it is known as the strong nuclear force. This force is not directly perceivable in the everyday world, of course, but the sun and the stars depend on it to sustain their energy. In that sense, we might note, the earth and all its life have always depended on nuclear energy.

The forces we are most strongly aware of are gravity, and the electromagnetic force. The electromagnetic force is 10^{38} times stronger than gravity. The nuclear force is in turn a thousand times stronger than the electromagnetic force. But whereas gravity and the electromagnetic force are long range forces that reach out across space, the strong nuclear force is extremely short in its range. The strong force is much more like two pieces of adhesive tape sticking together—as you bring them closer and closer, they experience no force until they are almost in contact, at which point they stick very strongly. This force binds the nucleons and confines them in the nucleus into a tiny boiling knot.

We should mention, in the interest of completeness, that there is another, fourth, force, the weak nuclear force, which we do not experience at all in daily life (and which may yet turn out to be a hitherto unsuspected aspect of electromagnetism). The weak force is only a ten million millionth as strong as the strong force and it controls the way that certain particles such as neutrons, if left undisturbed, spontaneously disintegrate or decay, into lighter particles. We will hear more later about the weak force.

If we were to list at this stage what we have found to be constant, what would we have? Some forces: the strong force and the electromagnetic and gravitational forces, and weak forces, all of which seem fundamental; and some particles: the proton, neutron, electron and photon. These particles are all in some sense fundamental. But they are not the only candidates for that distinction, nor are they themselves without structure. In fact current physics is much concerned with unravelling this structure: the proton and neutron especially behave as though they are 'made of something', maybe even 'made of something else'! But this is not all. The observations and experiments which over the last half century have increasingly confirmed the existence of these particles and brought us closer to understanding the forces, have also revealed, at that subtle level we keep speaking about, something else which is fundamentally constant. This constancy lies within the *behaviour* of the particles, their behaviour as they interact. This underlying permanence is in the *rules* by which energy is exchanged and by which, among other things, momentum is exchanged in transactions between particles.

You may notice a subtle shift here. We are now speaking about the transactions rather than the things. We are concerned with the dynamics of what happens rather than the characteristics of the particles as objects. We begin to be interested in *resonance* and *proportion,* i.e. the relationship between things rather than the things themselves.

If we have absorbed any of this then some of the preconceptions we are locked into, our waking sleep if you like, may have begun to dissolve. And questions may be arising about the validity of this science. Are the particles real? Is light really both waves and particles? Is all this really fact, or is it just mindfluff?

If we don't find it easy to accept this as fact it's not surprising. The relative comfort of our earthly life conceals from us the realization that on the cosmic scale there are a number of processes and reactions which are statistically more normal than the apparent solidity of the congealed energy that we are, and which we inhabit. It's a part of this geocentric shortsightedness that we forget that the universe is not principally composed of planets. It's mostly just space. (But not, as you might think, empty space, because it's everywhere full of starlight—photons, and a never-ending flux of other particles, the cosmic radiation.) The predominant objects in it by far are stars, mostly much bigger than our sun. In the sun and stars energy is far from being congealed into structure, and arises in concentrations unlike (until the first H-bomb) anything experienced on earth. Identifying what is constant and unchanging, i.e. fundamentally Real, in the matter found on the Earth is one thing; extending this to include the internal processes of the stars is another thing altogether. But unless we do extend it, of what value is the research?

To do so means making yet another, and this time perhaps a yet more subtle, leap into the dark. The darkness of an inner space much smaller and more violent in its basic dimensions that what we have seen so far, and yet amazingly and beautifully in proportion, it seems, with the huge vastness of the equally impenetrable outer space.

If we still feel that the particles remain somewhat fanciful to us, we might reflect on the fact that if we extend our hand in front of us, then on average (at sea level) one highly energetic particle, accelerated to almost the speed of light deep in space, goes through our outstretched hand each second

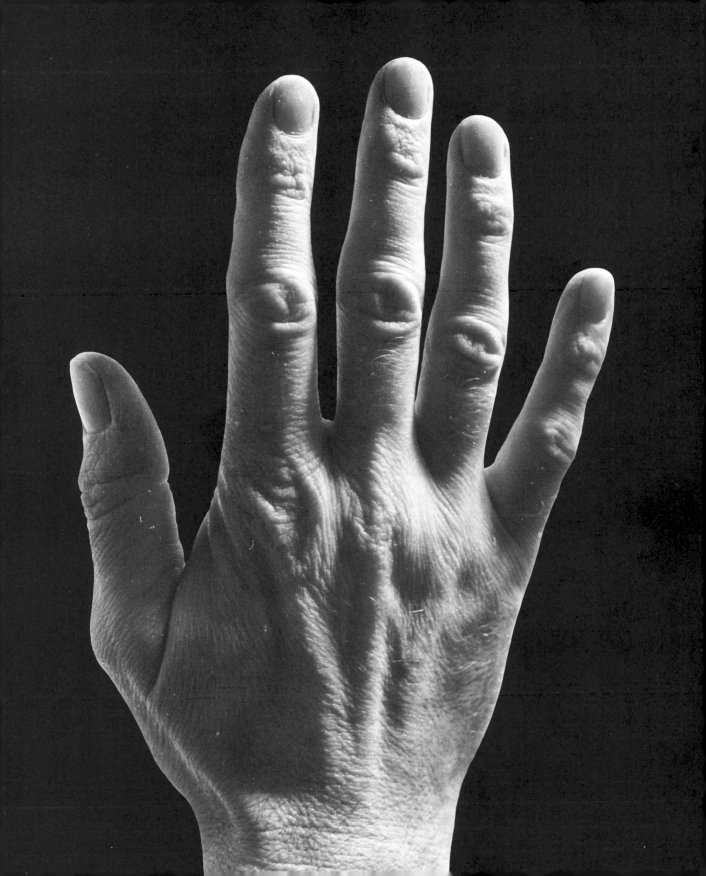

Size Look at a drop of water 6 mm across. Magnify it 2000 times so that it becomes 12 metres across. Here and there things the size of footballs are swimming about, single-celled creatures called paramecium. Magnify the drop another 2000 times; now it's 24 km across. It begins to look like a huge crowd of people seen from a very long way away. Only they're not people but molecules magnified to be as big as this full stop.

Again magnify the drop, this time 250 times; now it's 6000 km across. Now we can see that each molecule of water is an oxygen atom with two atoms of hydrogen tied to it. At this magnification of a 1000 million they look about as big as footballs.

Take one atom and enlarge it until *it* is 100 metres across. At its centre, only barely visible, is the nucleus, about the size of a grain of salt.

Energy If we reckon as a unit of energy the amount of energy it takes *per molecule* to break apart a salt crystal (of a block of salt), it would take a million similar units of energy to break the salt molecule apart into separate sodium and chloride atoms. And it would take a hundred million million of the same units of energy to break one nucleon from a sodium or chloride or any other nucleus. Ten thousand million million of the same units of energy *weigh* as much as a single proton (and are called a GeV or a thousand million electron volts).

Scale This salt crystal from the author's table salt (right), contains about a thousand million million million atoms. This book is about 250 million atoms thick (i.e. about 25 mm). If an apple was enlarged to the size of the earth, an atom in the apple would be about the size of the original apple.

Velocity Light crosses a nucleus in 10^{-24} seconds (a millionth of a millionth of a millionth of a millionth of a second). Light crosses an atom in 10^{-19} seconds (a tenth of a millionth of a millionth of a millionth of a second). Light crosses this page in a little under 10^{-9} seconds (a thousandth of a millionth of a second).

Number

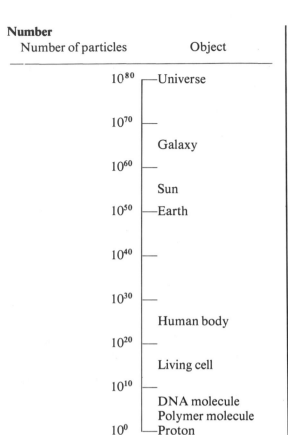

Number of particles	Object
10^{80}	Universe
10^{70}	
	Galaxy
10^{60}	
	Sun
10^{50}	Earth
10^{40}	
10^{30}	
	Human body
10^{20}	
	Living cell
10^{10}	
	DNA molecule
	Polymer molecule
10^{0}	Proton

Proportion Some idea, not just of the staggering scale of the universe, but also possibly of the profound relationships which are manifest in it, may be found in a group of relationships involving the very large number 10^{38} (one with thirty-eight noughts after it). Physicists have a very good idea of the size of the nucleus; it's about 10^{-13} cm in diameter. They also, through astronomy, have quite a good idea of the extent of the universe. It seems that the diameter of the universe is 10^{38} times the diameter of the nucleus. The physicist's basic unit of time (he doesn't have much use for earthbound seconds) is the time it takes light to cross the nucleus! The age of the universe, arrived at by other means, appears to be 10^{38} times the time it takes light to cross the nucleus. If this were not enough, it seems also that the relative strength of attraction of the electromagnetic force and gravity is similarly proportioned: i.e. attractive electromagnetism is 10^{38} times stronger than gravity. And finally it has been calculated that there are just $(10^{38}) \times (10^{38})$ protons in the universe.

This is what we mean when we speak about the cosmic scale. As we go on deeper into what high-energy physics reveals, it's good to remind ourselves that if this kind of relatedness on this kind of scale does operate, perhaps we would find, if we are able to look, a similar relatedness at all levels of our experience.

5. No-Thingness

To be able to enter into a substantial relationship with the kind of situation outlined in the previous chapters requires of the reader, as it has required of the physicist, something of a break with the past. It requires a major step into abstraction.

The first barrier we should recognize as a problem that isn't going to go away, is that of the mathematics. Don't disappear. There isn't going to be any mathematics; a few numbers later on maybe, but no maths as such. This is only a mixed blessing because the advantage of mathematics that is denied to us, is its precision. All physics proceeds by analogy; as we've seen, the labelling process of language itself is intrinsically analogous, but mathematical analogues may fit exactly the behaviour of some aspect of nature. For instance Maxwell's equations precisely describe the forces of electromagnetism found in electric motors and radio waves.

$$\nabla \cdot E = \frac{\rho}{\epsilon_0} \qquad\qquad \nabla \cdot B = 0$$

$$\nabla \times E = -\frac{\partial B}{\partial t} \qquad c^2 \nabla \times B = \frac{\partial E}{\partial t} + \frac{j}{\epsilon_0}$$

So we must accept that to an extent there is a barrier here. But if we are smart it may help to focus our attention on essentials, on what physics *reveals* rather than what it *is*. Take all this as a signpost to more acute awareness, not a teacher text; don't read for facts or precision, read for orientation.

Let's plunge ahead and enter the wholly new inner world that physics has brought to light in the first half of this century. First remind yourself of our earlier ideas about what constitutes a particle. Think of just space. Empty space. Now think of it as potentially capable of forming pinched-up centres of energy. Let's for the moment think of them as protons. Physicists label a piece of space that can be pinched up to form protons *a proton field*.

Now back to the space again, the same—empty—space. Now think of it again as potentially capable of being pinched up into tiny centres of energy. Only this time the centres of energy move in the space. At 186,000 miles per second. They can ONLY move at this speed. What we have imagined is a 'photon field', or more properly and scientifically *an electromagnetic field*. Our piece of space can also be thought of in its imaginary dormant state as potentially able to be manifest as electrons and neutrons and protons too. Each of these fields is held to be present in the same piece of space, throughout *all of* space. Of course we ought to add the gravitational field as well; this is also believed to have a particle, the graviton, but it has not so far been detected.

Does this now seem a simpler picture than the atomic portrait of our room as a soup of particles? At one level of course the atoms are really there; but at this more subtle level it is fields that are 'there'—fields which are always present, and which have constant properties, but which may be temporarily manifest as particles. Or, to put it another way, the fields endow the space with constant properties.

This is a point where we need to make our mental model of what happens in sub-atomic events a little more sophisticated than before, to incorporate fully the idea that, since the particles have proved highly transient, it's their *behaviour* that matters. We have to move towards the idea that the phenomena of the world are the result of only four kinds of *interaction* between the many fundamental fields, and *nothing else*. The key word is *interaction*. It may seem like just a fad to give it so much emphasis, but it really is a very practical concern, because thinking of sub-atomic events as interactions greatly simplifies them.

Atoms for instance are thought of as the result of an *electromagnetic interaction* between the proton, neutron and electron fields. The nucleus however is held to be the result of a *strong interaction* between the proton and neutron fields. The neutron when it decays into a proton and two other par-

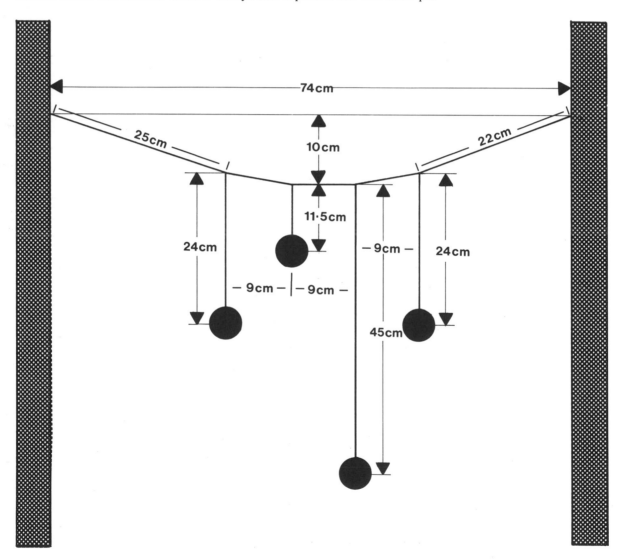

A model which may help us to understand what is implied by the emphasis in this chapter on *interaction* can easily be made up from simple materials such as wooden balls and string. To operate the model, set the righthand wooden ball swinging (backwards and forwards as we look at the photograph and diagram in the version pictured here). The swinging ball sets the centre two balls moving, but irregularly. However, the motion of the first ball is gradually transmitted to the other ball which is on a string of the same length, the one on the left. In fact after about 14 swings, the first ball stops and the one on the left is swinging almost as far as the first one. Then the process is gradually reversed until the first ball is moving again and the one on the left has stopped, and so on. The analogy with particle interactions runs like this. Like pendulums, particles are identical vibrating systems which can resonate with each other. The horizontal string coupling the particles represents the relative strength of the interaction (for gravity a thread, for the strong interaction a rope, etc.). Try it, and see for yourself

ticles, is held to have decayed through a *weak interaction*. As you sit in your chair, all the atoms of your body are having *a gravitational interaction* with all other atoms everywhere in the universe.

Not all the interactions have the same strength (or time scales), of course, nor do all particles take part in all interactions. All particles have the gravitational interaction, which is weaker than the strong interaction by 10^{40}. Most have the weak interaction, which is 10^5 times weaker than the strong interaction; some particles have the electromagnetic interaction, which is only 10^2 times weaker than the strong interaction. By now a large number of particles have been found which, like the proton, take part in the strong interaction.

Physics is interested now in nature's actions. What can happen. What can't happen. Again we meet the idea of nature not as an aggregation of 'things' but as an 'event'. As we shall see, the rules that govern what can and can't happen seem to be turning out to be more substantial than the objects they make comprehensible.

Einstein is reputed to have once said that 'the most incomprehensible thing about the universe is that it is comprehensible'. Does it begin to seem comprehensible? Try to think of it as a dance for which the choreography is slowly being uncovered. The dance is pretty but transient; but the choreography from which it arises looks to be more subtle, and contains within hints of something permanent and unchanging.

The subtlety of this way of coming to know nature lies in its simplicity. There may be many different fields, each with a potential of unmanifest particles, and these many fields may be complex, but the *behaviour* of the particles is much simpler. They appear to interact with each other in only four ways. From now on, questions about what they are *made of* give way to enquiries about their *behaviour*.

This is another crucial point in our penetration of physics. Before we go on, let's be sure that we've got it straight. Up to now we've seen that we are, the world is, or can be thought of as an accretion of molecules, which are in turn accretions of atoms. Atoms are structural forms of particles and forces; and particles are local perturbations of space. Each of these perturbations of space may arise from and return to a fundamental field corresponding to each different particle. Particles are to be considered transitory and impermanent. Our room has become more than slightly alarming. Can we survive its uncertainties? Sir Arthur Eddington wondered about that too:

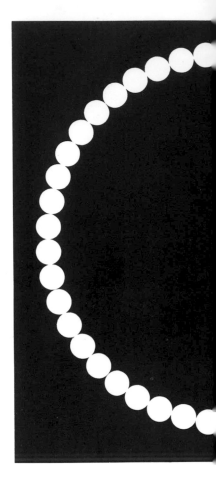

> I am standing on the threshold about to enter a room. It's a complicated business. In the first place I must shove against an atmosphere pressing with a force of fourteen pounds on every square inch of my body. I must make sure of landing on a plank travelling at twenty miles a second round the sun—a fraction of a second too early or too late, and the plank would be miles away. I must do this whilst hanging head down from a round planet, head outward in space, and with a wind of aether blowing through every interstice of my body. The plank has no solidity of substance. To step on it is like stepping on a swarm of flies. Shall I not slip through? No, if I make the venture one of the flies hits me and gives me a boost up again; I fall again and am knocked upwards by another fly; and so on.

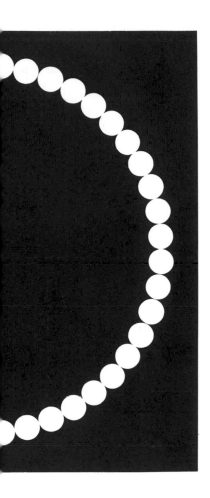

But within this shifting cloud of possibilities (or as we'll see shortly, probabilities) there is permanence. Immutability of essential behaviour. Perfection.

So let's move on, if you are not too dizzied by the pace, to consider some further properties of the fundamental fields. Let's go back to our piece of empty space. Think of it again as just space. Now think of it as pinched up. Into a particle. How big is your particle? A point? Smaller? Larger? All right let's leave the foolishness! In fact it's a characteristic of the fields we're speaking about that there are strict limitations on smallness and bigness. In this beyond-the-senses-world *only certain sizes are allowed*. Energy only pinches up in certain definite sizes. If you keep in mind the wave aspect of particles, then think of the wavy pattern round the edge of a plate, you may get the sense of why only certain sizes occur. In nature of course, the waves are three dimensional, not flat, but they too must join up when they meet round the other side, just as the wave pattern on a plate does. On the plate the waves divide the edge in whole numbers; there can be one, two, three, four, five . . . sixteen, seventeen, eighteen, nineteen, etc., etc. waves but not five-and-a-half or seven-and-two-thirds waves.

This is a key factor in understanding the behaviour of matter at these very tiny dimensions. It means that quantities are fixed and constant. It means, for instance, that if the proton field is manifest as a proton, the proton will always be the same size and mass—wherever or whenever it appears. Which in practice means that all protons are absolutely indistinguishable, since they are all absolutely identical. It's almost as though when physicists detect a proton it's always the same one popping up in different places. The same is true for all other particles and for all other fields.

If we pause for a moment we may be able to see just how extraordinary this is. We may be able to see that if it were otherwise . . . there would perhaps be no otherwise. If there were even the slightest deviation from the absolute perfection of dimension of each particle, coherent structure would be impossible. The atom would not occur, there could be no coherence of compounds nor formation of crystals, therefore no metals. As it happens it would seem that the universe has been made in the only possible way; that on the cosmic scale everything that can happen does happen – the universe we have is perhaps the only one that can exist.

Again we have found simplicity. Many fields and many particles, but only certain sizes are 'allowed'. There is an immutability of dimension. This immutability of size means that form pervades the otherwise apparently diffuse nature of the fundamental fields. The fields can only be manifest in certain discrete, i.e. definite, quantities (the official science name is 'quanta') of mass and energy. This in turn means that the process of energy transfer and excitation in atoms and in the particles themselves, while superficially a disorderly turmoil of movement and change, is in fact a very highly ordered energy flux. So that electricity flows in wires when the quanta-ty of electric charge in a metal overflows as free electrons; light quanta may be emitted when the quantity of energy in atoms overflows the allowed value. And so on. Throughout the whole universe the rich baroque texture of energy transactions is interfused by a precise order, the quantum rules, which allows nothing to escape, nothing to go spare, no exceptions.

A further simplification of the particle fields remains tantalizingly just over the horizon of the physics imagination. Could it be that all the many particle fields are really different 'views' of the same single field? Could it be that the four interactions are also somewhat arbitrarily separate? Could they not be incomplete pictures of a single underlying something? Is it really more likely that they are *intrinsically* separate?

The quantum field is a stage. On it the cosmic ballet of existence is played out. However transitory or permanent the dance which its surface presents to us, we may take this display as being a creation which arises from it and ultimately returns to it. And beside which it may be that all the calculations and measurements of physics are merely descriptions of the flickering of a celestial candle flame whose Reality we are unable or unwilling to experience.

You might well feel inclined to ask, and people have, how do we know all this? *Do* we really know all this?

The answer is not exactly. Particle physics deals with very small dimensions and with very large numbers of particles and because of this all measurements are statistical in their nature. In fact it seems to have been established beyond question that there must always be an element of doubt in measurements at these tiny dimensions. But this doubt or uncertainty is itself a definite quantity. It can be included in the calculations. It means nevertheless that there is a fundamental limit to the certainty with which we can know where a particle is or where it's going. The uncertainty lies in the fact that one cannot know for certain BOTH the speed and the position of the particles. Only one or the other. The more accurately we wish to know one, the more we must sacrifice of the accuracy with which we could have measured the other. The very act of *measuring* the one will alter the result of what we might have found by measuring the other.

Quantum physics requires the acknowledgement that the observer is not just a passive spectator, but an active participant in the universe whose laws he is trying to understand. The reason for this is that the particle we use to *look with* disturbs the particle *looked at*. We can never be sure what could have happened to the particle after the point where it was hit. In consequence, the closer we wish to look at the sub-atomic world the more we can deal only with what is probable or relatively certain.

There is nothing intrinsically difficult to grasp in the notion of probability. Suppose for instance we wish to know how many people prefer tea to coffee in Britain. If we ask a hundred people, we may be able to guess with a fair amount of certainty how many people in the whole population prefer tea to coffee. If we ask a million, then it gets to be exact *enough*. Similarly, insurance premiums are based on the statistical probability that certain classes of people will either have more or less car accidents, or will or will not die at certain ages. Its the same with betting. Profits in gambling are based on the certain knowledge that the probability of the odds over a long period will favour the bookmaker. There may be day to day fluctuations but the average remains steady.

The notion of probability is closer to us than we might have thought, but

The tendency to think of ourselves and the world as built of molecular bricks and mortar is a difficult mental habit to break. Yet we should keep firmly in mind that the cosmic ballet is manifest in the form of wave-like particles, interacting with each other in only four ways. (A useful rule of thumb is to think of energy propagating as waves and being absorbed and emitted as particles.) No direct image is entirely satisfactory, since we are unable to 'see' three dimensional waves, but water waves can lead us to some sense of the rippling irridescent quality that underlies the mundane surfaces of our surroundings

Overleaf: surprising though it may seem, a truly random series of numbers or arrangements of objects constitutes a very highly recognizable type of order, and a whole branch of mathematics deals with the statistics of probabilities associated with the fluctuations of large numbers of events. For instance, the distribution of leaves falling from a tree is predictable within certain limits. Knowledge about particle interactions has a similar relative precision.

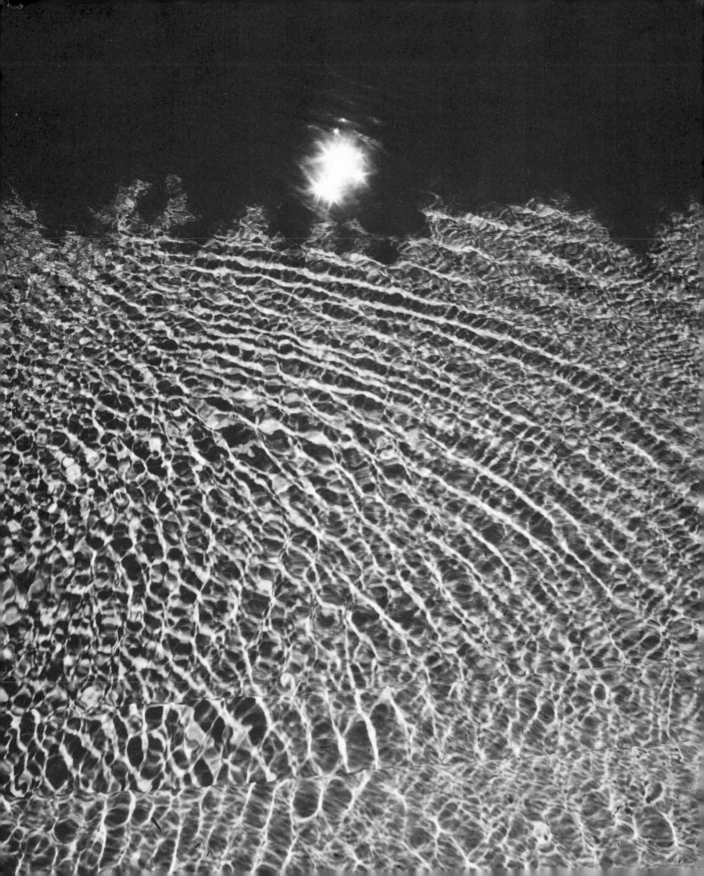

Below: gambling is founded on the exact knowledge that if racing is genuinely competitive, over a period of time the odds will favour the bookmaker by a certain definable percentage

quantum physics builds probability into the very core of physics law. The complementarity between position and speed (more properly momentum) means that, about one at least of these aspects of the description of motion, only a probability can be determined. Certainty has dissolved into a precise calculation of chance. At the sub-atomic level all knowledge is eventually of this kind.

Below: gambling is founded on the exact knowledge that if racing is genuinely competitive, over a period of time the odds will favour the bookmaker by a certain definable percentage

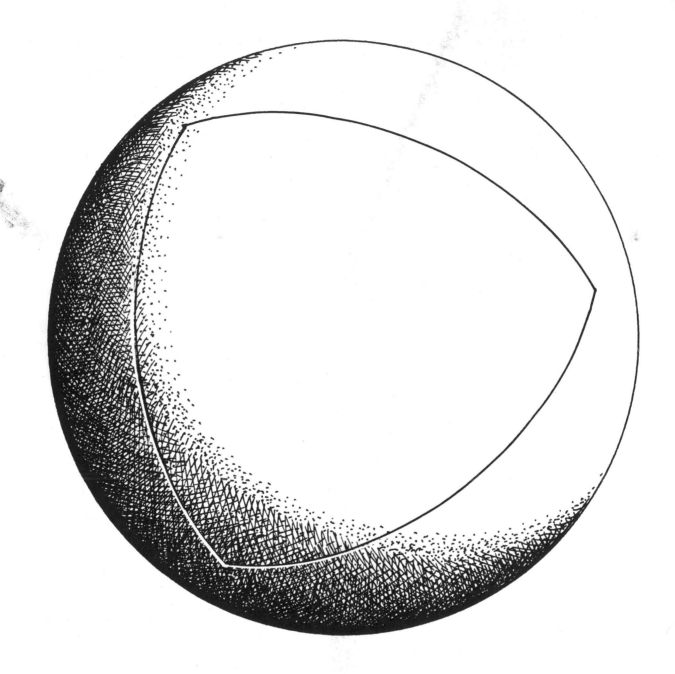

This is the end of the beginning. This is where we have to leave aside the study of our room. The events and occurrences of hi-energy physics are about to fall on you. But perhaps one further general piece might be appropriate. You see, in speaking about a subject such as the present one, it is often essential to 'know-it-all-at-once'—coming to know a bit here, depends on a prior knowledge of the something else which, in turn, is dependent, etc. And we are in that corner here, believe me!

The last general topic that needs to be brought into view involves various insights which derive from the absolute nature of the velocity of light. In a word: relativity. Much that we are about to encounter in later chapters hinges on some appreciation of the intrinsic connectedness of the constituents of the universe. Relativity theory, originated by Einstein, remains the most beautiful and most comprehensive of physics' simplicities.

As we move about in the space of our own personal territories, taking time to devote our energies to whatever needs or whims might come to mind, we are not usually aware of the specific relations between time, energy and space. On the scale of human actions in the living room we do not notice their connectedness. But at the normal scale of the universe, whether astronomical or microscopic, the connectedness cannot be ignored. One of physics' major enterprises has been to uncover highly coherent relationships beween time, space and energy. They are intrinsically related.

These relationships, so far as we need to speak of them here, turn on the determination or discovery of the velocity at which light is propagated through space. This as we've heard is 186,000 miles per second. But the deep significance of this very high velocity for physics is that it appears to be *absolute,* i.e. it is fixed and unalterable. And not only that, but it is an absolute limit to any velocity at all by anything. Or so it appears at this point in time. This constancy is intrinsic: it turns out that, however or wherever it is detected, light moves at this velocity. Even if you are moving at half the speed of light towards a source of light, you will find, if you measure its speed, that it is the same 186,000 miles per second velocity. Not only that, but even if you are moving away from a source of light at 93,000 m.p.s., and measure the incoming light, it will still be 186,000 m.p.s., not half that number. Not 186,000 m.p.s. minus 93,000 m.p.s. Many people have thought long and hard about that. You think about it. No, don't just think about it, try to realize what it means.

Can you see what the consequences are? The clue is that everything in the relationship between light and the detector of the light is intrinsically connected. In fact we could then guess that if the observer travelling towards a source of light at half the speed of light detects that the incoming light is at its usual velocity (i.e. their speeds were not added), then you would expect some change to have occurred to the *observer*. And this is the case. As the observer accelerates in velocity, there is a corresponding increase in his *massiveness* compared with his previous mass before he started moving. Though he would have no way of detecting it himself, someone else might. Of course at slow speeds, the increase in mass is slight. For instance if a two-ton spacecraft travels at a velocity of 25,000 miles per hour—the energy of motion is only equivalent to the rest energy of one hundredth of a gramme. But at very high speeds, i.e. close to the speed of light, the mass increase can be considerable.

We might then ask what else happens to this energy that went into accelerating the observer—this energy that has turned into greater massiveness in the fast-moving observer? Does it just continue? Can it disappear? What happens to it? What would happen if the fast moving observer was made to collide with a fixed target? Would the energy just slip off him?

You're a creature living on a surface and you are not sure whether it's curved or flat. How would you decide? One way would be to draw a triangle and measure the sum of the angles of the corners. If they add up to 180° the surface is flat. If not, it's curved. If in effect we measure the angles of triangles in space and we find that they do not add up to the expected figure, we may conclude that space itself has curvature, which is a research finding that is now well established

Of course there is more to relativity than can be even hinted at here. The key ideas turn on several aspects of the intrinsic connectedness of the universe—that, for instance there is no such thing as a fixed space and that there is nothing in the universe we can be certain is stationary. For all practical purposes everything is adrift in space at different speeds, and so if we make an observation we have to take into account our own frame of reference—Are we moving? In what direction? How fast? Etc. The unifying factor in relativity theory is that, regardless of his motion, an observer will always obtain the same answer if he measures the speed of light. This in fact is reflected in a new geometry appropriate to movement through space.

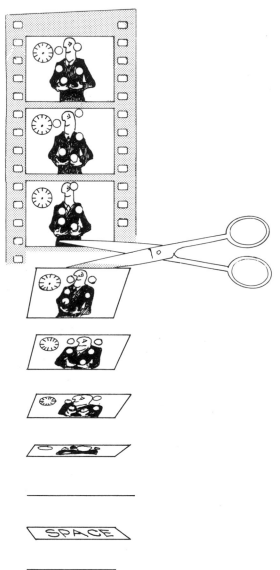

Frames of reference

Events happen in space and in time, but space and time are not absolute or intrinsically separate. Relativity unifies space with time. This has profound consequences for the understanding of phenomena, especially in situations involving high speeds, such as in atomic and sub-nuclear physics, and also astrophysics.

Think of a man juggling. From one point of view, we can picture his actions as a sequence of events in space arranged in time—like successive frames in a film. Think of these frames, each representing an instant snapshot of space, arranged in a stack of slices. Imagine now the stack as a continuous piece of space and time—space-time. What of the inside of the piece of space-time?

Think now of one of the balls. Its image inside the stack is like a sausage, a zig-zagging spiral made of a pile of slices.

Think of the block of space-time being sliced in different ways. Each different slicing makes a different separation of space and time—corresponding to a different point of view or frame of reference. Think of these slices reconstituted to form a film again. The same events of the man juggling could look very different from this new point of view. Indeed, if one of the balls had been moving with constant speed in a straight line, one parallel slicing could make the ball seem stationary, while everything else moved around! Since the balls were not in fact moving with uniform motion, but accelerating and decelerating as they rose and fell to the hands, to slice space-time so as to make one of them appear stationary, needs a *curved slice* (which is what Einstein's general relativity is principally about)

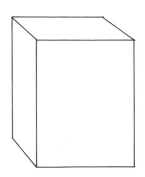

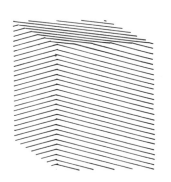

The answer is that this is exactly what does happen, with far-reaching effects for the whole of physics. We'll come to this again in the next chapters.

These relationships have a number of other interesting, if contrary to common-sense, consequences. If velocity, time and space are inter-connected, a new description of the universe-wide equilibrium of forces, structures and pattern must be written. Previously, space had been considered to be uniform and consistent; in the space of Euclid the angles of the corners of a triangle always added up to 180°. But now science has come to accept that space itself is warped, or has curvature. For instance, in space warped by the presence of gravity, the planetary orbits can be happily thought of as straight lines in curved space-time. And a spaceman who sets off ← this way to find the edge of the universe will return, if he returns, ← this way.

Another way of putting it is to say that mass-energy tells space-time how to curve, and curved space-time tells mass-energy how to move.

Can you see the beginnings of a deeper simplicity in all this? It's a very grand scheme that physics has uncovered and it is very daunting to come to it *en masse* like this; but remember, also, it is simple. Our minds are just not used to experiencing this level of connectedness. There are three key ideas in this chapter. One, the idea of a number of fundamental fields, each of which is a property of space itself, and each of which may be thought of as a property of one single, simple, underlying quantum field. Secondly, the notion that the particles that may arise in these fields and which interact in only four ways, have definite limits to their sizes, they exist as quanta. Thirdly, just as the four interactions may be thought of as one interaction seen from different points of view, so also there is a unity of relations between time, space and energy.

Take it lightly. But take it to heart. This is the fabric of our being.

This model attempts to suggest
the way that space is warped by
the gravitational field of a
massive object. Everything
(including ourselves) has some
effect of this kind on the whole of
the rest of space

6. Collisions

How do we know all this? Or indeed, do we know all this? It is difficult for those outside science to accept the reality of the kinds of discoveries we've been hearing about. If you have no experience of it on a day-to-day basis, the scientific method can certainly seem very mysterious. As usual it's really much more simple than it seems. Here is one way of experiencing something of the situation that the physicist is in.

You will need someone to help you. Go and find a friend. Tell him or her you want him to guide you through a little experiment. Now friend, your friend with the book here wants you to help him with a little demonstration of something that the book needs to make clear before he goes on to the next section. This is what I want you to do. Get your friend to close his eyes. All right? Eyes closed? Now can you think of a small room somewhere nearby with a door? Preferably one which he's unfamiliar with? Turn him round a few times, or lead him by a confusing route to the room. It's important that he doesn't know what the function of the room is. (A good way to ensure this is to give him some ear plugs to restrict his hearing or to have him wear a pair of earphones. He should still be able to hear your instructions of course!) All right now, eyes closed? Off you go. Take him gently by the hand and lead him to the room you've chosen. Make sure he doesn't cheat! Next give him a pencil or ball pen or a short piece of wood or something about that sort of size. Now, you are outside the room and you are quite sure he doesn't know where it is or what it is (ideally it would be a small washroom or a broom closet or a bicycle shed). Tell him to go into the room keeping his eyes shut and see if he can find out what the room is, what its function is. But the only thing he's allowed to do is to tap with the end of the pencil. All right let him start. After he's got into it tell him that he's only doing what the physicist does. He'll know what you mean. The physicist works similarly in the dark, he can never 'see' what it is that's he's working with, he has to guess, just as your friend is guessing, but guess intelligently, knowing that he can't be altogether certain.

Let your friend tap around with his pencil for a minute or so, then ask him to come out for a few seconds. Ask him if he thinks he knows what the room is (he must keep his eyes closed of course). Try to get him to form an opinion of what the room is. He should then be able to think of something that would confirm what the room is. If he can, let him go in to search for it. For instance, if it was a broom cupboard, he might come out and observe that he thought it was a bathroom and then decide to go back in to establish taps as confirmation of that.

Opposite: a spray of photons emerges from the sun and transfers its energy to the atoms in the glass molecules of the window . . . (see text)

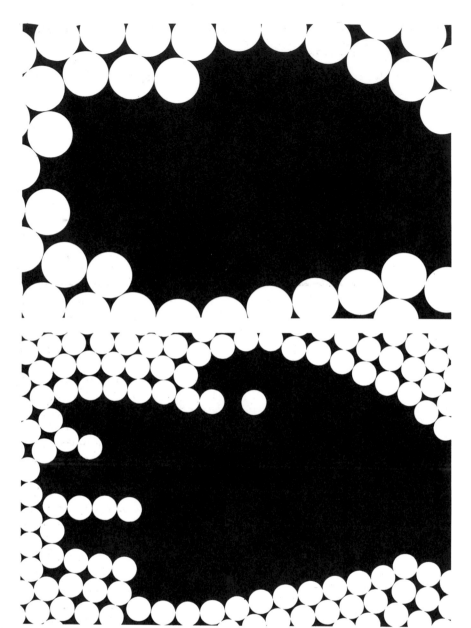

The problem facing microscopy is analogous to that of making an image or a mould using different sized balls. Footballs would give no image of a hand. Marbles only a little better. Sand would be excellent, but it would take something as fine as flour to make a precise image.
A photon is effectively many thousands of times bigger than an electron. This is the reason why electron microscopy can see much finer detail than microscopes using visible light (see overleaf).

The same sort of game could of course be played by getting a friend to select a number of objects and lay them out on a table, and then with your eyes closed, but still only by hitting them with a pencil, you try to identify the objects. What should be clear from this is that the information is constructed from a series of *collisions*.

This portrait of the scientific community, and in particular the physics community at work, may strike us as hilarious, if not altogether unattractive, but the scenario of the visit to the broom cupboard may nevertheless be taken as a

rather precise model of the way that science creates information about the world.

A physicist is just as certain about what a proton is as you were about the room you were just in. What you may not have guessed from the analogy is that any kind of looking means making collisions.

If we momentarily emphasize the particle aspect of the interactions of our environment then it's clear that all looking, all kinds of contact at all levels, involves collisions of one sort or another. A spray of photons emerges from the sun, transfers its energy to the atoms in the glass molecules of the window. These are called glass because they transmit the energy of the photons, so that on the other side of the glass more photons very similar in wavelength to those from the sun, are emitted. These emitted photons cross the room, collide with the atoms in the paper of the page and then, because they have raised the energy level of the atoms in the page to an intolerable level, other photons are emitted from the atoms to travel up and out from the page in all directions. A hail of particles may then impinge on the eye. One photon hitting the visual purple dye of the retina can be absorbed and will make the vibrating molecule trigger off an electro-chemical process which we call a nerve impulse, sending a signal to the visual cortex at the back of the brain, where it is 'perceived'.

However we 'look', a similar process is enjoyed by the particles we choose to work for us. If we can get this clear then we will see that microscopy is a very refined version of this process.

At the level of the unaided eyes, photons of visible light from the sun collide with the petals of a flower. They overload the energy capacity of the atoms of the petals, and in consequence more photons are emitted. Some of these are allowed to penetrate the glass of a camera lens and subsequently to fall on a piece of photographic film. The colour of course comes from the flower atoms' ability to absorb selectively some frequencies of the white light (which is a mixture of all frequencies) falling on it, so emphasizing others, in this case red.

The next stage might be to look at something through a conventional light microscope. Here photons of visible light colliding with a tiny area of a specimen are focused and brought to the eye by a lens. But the photons are quite large compared with the details in the subject—it is like making an image of a hand using, say, footballs. To see more detail we must make our collisions with something smaller. The next smallest particles available means using electrons in an electron microscope. Here again the sequence is similar to the light microscope, except that now we are making an image of the hand equivalent, say, to that we could make using tennis balls. The finest details are still fuzzy.

Under certain conditions atoms can be made to behave as if they were much smaller than electrons. In a field ion microscope metal crystals can be bombarded by other atoms and as these atoms bounce off they can be detected in such a way as to make a direct image of the arrangement of atoms in the crystal. But by now the collisions have become so energetic that the metal crystal atoms can only just be kept from boiling off. The act of looking is almost disintegrating the object being looked at.

But what happens if we want to look at, to make an image of, the separate parts of the nucleus: the nucleons, protons and neutrons? To 'look' at the components of the nucleus we have to literally *break it apart*. This can be done if the nucleus is hit by particles that are energetic enough to overcome the strong force that is gluing the nucleus together. But we have reached the point where the act of looking irrevocably changes the thing looked at. Nevertheless, the information gained in the collisions is still reproducible; a carbon atom will always turn out to have six protons.

But the really big jump that puts physics into a whole new and unprecedented situation comes when what is 'looked' at is the nucleon itself. And especially the proton. Here the collision has to be a thousand times more energetic than that to break apart the nucleus. What happens now? This huge amount of energy (which is remember, needed even to make contact with a proton), not only changes what is being looked at, but in a very real sense, *creates* new things.

And not only that, but each 'look', each collision, creates *different* things. This unique and extraordinary situation sets the sub-atomic landscape apart from the every-day street world, and physics is still trying to come to terms with it. For this reason, and because it dominates current physics theory and experimentation, it forms the core of the next few chapters.

It introduces us to the mechanism of high-energy physics and to the idea that, though they do not look it, the vast installations that operate in many parts of the world spending upwards of tens of millions of pounds sterling a year are all merely very sophisticated kinds of microscopes. This is a fact which has tended to become more and more obscured as the installations have grown in size, complexity and cost. So, please remember, from now on, whatever it may sound or seem like, what we are speaking about are dis-

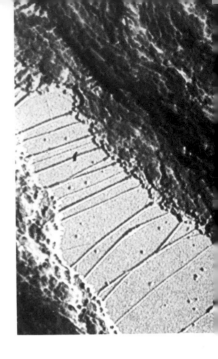

Below: low magnification, ×125, picture of lead-bismuth-tin-antimony alloy. Discussion of microscopy here helps to emphasize that what follows in later chapters is, despite appearances, still very much a form of microscopy

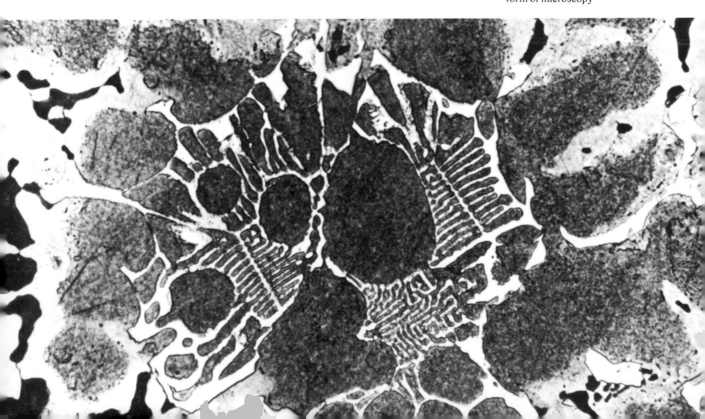

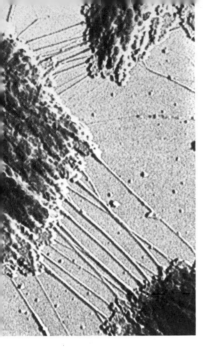

Above: fine detail seen by
electron microscopy—the
links between polymer crystals
in polyethylene, at 40,000
magnification

coveries and revelations of the current state of the art of microscopy. High-energy version.

The next chapter gives a brief visual taste of one of these vast installations, that of CERN, the European Centre for Nuclear Research which is near Geneva in Switzerland. Here again the surface conceals the function of the machinery almost entirely, and we may find it useful to look first at a practical model which can rather completely simulate, on a smaller scale, what the whole of a place like CERN is up to.

This demonstration, like the earlier ones, may need an accomplice to make it work best; but it's worth considering without one. Find a room somewhere in which it's O.K. to fire an air rifle, or if you're that rich, a bb gun. (You could set up the demo outside in a field of course.) You will need a rifle, either an air gun or ideally if it should be available, a fairground-shooting-gallery type gun which fires a stream of pellets on demand.

Obtain or make a large paper or card box, say, about a metre cube; it need have no top or bottom. Suspend the box so that its centre is at about shoulder height, and have your friend suspend in the centre of the box a heavy metal object about two or three inches in diameter. Ideally you should not know what it is. O.K.? Now, here's the thing. You are a physicist, the box contains what is believed to be a proton. Your task is to determine its shape, or get any information that you can about it. And all you have is the rifle loaded with pellets. The pellets are also protons, but high energy protons. See what you can find out about the proton by shooting at the white box (there's no need to be perverse about where the object is in the box, it should just be centred). Fire off a few rounds at the centre of the side facing you. Now if you go to the back of the white box what you will find is that, if your fire was accurate, there is the beginning of a shadow of the object in the centre, which will suggest its outline. There will also be a scattering of pellet holes over the back of the box which will tell you something more about the shape of the object. If you have the opportunity, build the set-up and try different shaped objects. It's surprising how exactly you can deduce what's inside.

The whole situation is a rather precise analogue of the function of a particle physics laboratory. The physicist seeks to determine the properties of the sub-atomic particles, to understand what they are. But the only way to make contact with sub-atomic particles is to hit them with other similar particles. A big laboratory such as CERN or Brookhaven is essentially a sophisticated array of just the four elements in our last demonstration. Physicists—particle accelerators (the gun) producing beams of particles (the stream of pellets)—targets (the hidden object) and detectors (the paper box around the target). The complexity of the laboratories comes from the need to work with beams of particles of prodigiously high energy and the need for very precise measurement of on the one hand, huge numbers of collisions (for statistical precision) and on the other, the extremely short time periods of some of the events being measured. All these requirements, high energy, billions of collisions and very short time-scales, have interacted together over the last fifteen years to produce an explosion in the size and cost of high-energy physics. This really is Big Science. Look at it.

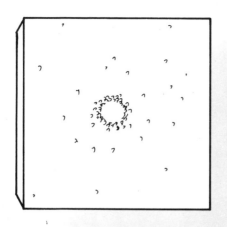

If you go to the back of the
white box (drawing inset) you
will find the beginning of a
shadow of the hidden object

Opposite: more interested perhaps in his machine time and the reliability of his soldered joints than in the philosophical implications of his work—the practical preoccupations of experimental physicists are reflected in this picture of the desk of a CERN experimenter, photographed just as it was found

Above: CERN is primarily a supplier of beams of accelerated particles. This beam is pulsing about every two seconds, each pulse supplying over three million million protons to an experiment

Below: the proton beam usually collides with a metal target, seen glowing (centre) from the impact of the protons

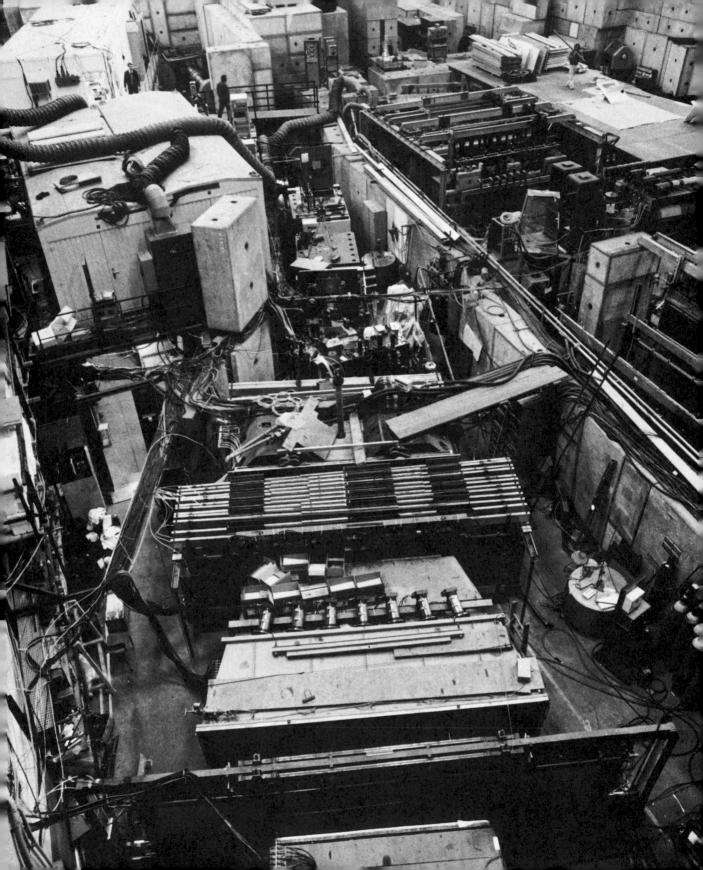

Left: some particles selected from the first collision collide again with a target (hidden top centre). The particles that come out of the collision are detected by the bank of spark chambers (bottom centre) the equivalent of the back of the box in the model on pp 116–17
Above and below: in another type of experiment two beams of particles travelling in the opposite direction in two intersecting rings are made to collide head on. In terms of the model on pp 116–17 this is like using as a target the bullets from another gun!

Prêt d'un sniffeur à l'Équipe
de Gargamelle — 17-09-70

8 AVRIL 1971

16.100.000 photos
ou
5313 Km de Film
ou
18,600 Km²
et
150.000.000.000 bulles ~
Sans les Doubles

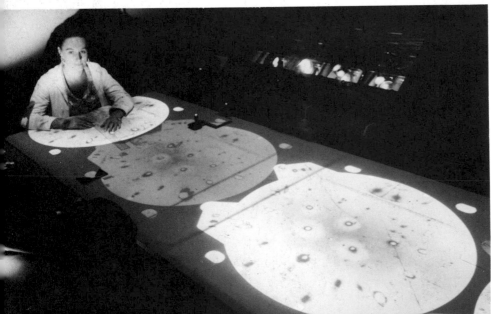

A major part of CERN's work lies in the analysis of the information obtained from experimental collisions

Top left: as long ago as 1971 Gargamelle, one of CERN's major detectors, had taken over 16 million photos on 5313km of film

Bottom left: with the assistance of a computer, each picture is measured to identify the particles and the interactions it reveals

Above: the explosion of data from this and other experiments is overwhelming, and is paralleled by the huge number of papers produced by physicists as they endeavour to understand the new world they are revealing

Right: a typical corridor interaction at CERN

$$K^+ \to \mu^+ + \nu + \pi^0$$

$$M = G \sin\theta \langle \pi | V_\alpha | K^+ \rangle \, \bar{u}_\mu \gamma_\alpha (1 + \gamma_5) u_\nu$$

$$e^{+ip \cdot x} \langle 0 | \partial_\alpha V_\alpha | K^+ \rangle = f'_\pm \quad 1_-$$

$$\hat{\theta}_{00} = \hat{\theta}_{00} + u_0 + c\, u_8 \qquad\qquad t = M^2 = (P_K - P_\pi)^2$$

$$m_\pi^2 \to 0, \quad c \xrightarrow{?} -\sqrt{2} \qquad t - \mu^2 \qquad \pm \mu^2 \qquad (p_\pi = 0)$$

$$G \sin\theta \, \langle vac | a_2 | K^+ \rangle \; const.$$

The new 300 GEV accelerator presently being built at CERN features a magnet ring (above) that is 8km in circumference (the circle on the aerial photograph below). The bigger the physics installations become, the more deeply they express confidence in the power of the mathematics (left) that both nourishes and feeds on them

8. The Cosmic Dance

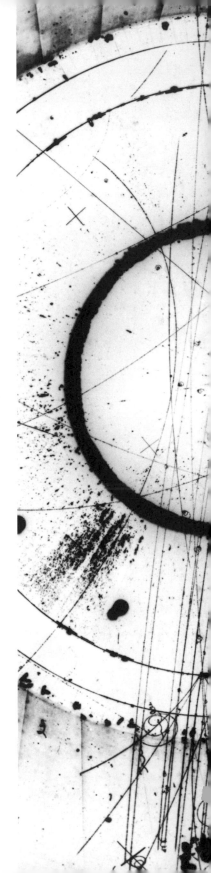

As was emphasized in Chapter 6, installations such as CERN are in effect huge microscopes. As we've also seen there is a permanent fundamental connection between the resolution, i.e. the sharpness, of a system for 'looking' and the energy it uses. The smaller the thing you wish to look at, the greater the energy that must be employed in the process of imaging it. As we've also seen, the gigantic town-size machinery of CERN is much simpler than it looks. A stream of particles is accelerated and is made to collide with a target, which is also, of course, made of particles. Detectors then allow data to be collected on the outcome of the collisions which, after analysis, can tell us what is 'there'. This sort of simplification does not, of course, do justice to the thousands of people who each contribute to the state of the art skills in supplying the beams of particles, creating the collision situations, and sifting the data; but here we are only concerned with what is revealed.

What does the microscope see? Of course we know by now it sees nothing directly, but we can infer that something like this happens. Some visual models are coming up, but first a word model. Think of yourself as a proton. A local perturbation. A pinched-up parcel of energy in an ocean of unmanifest quantum effects. There you are sitting quietly minding your own business, occupied with the work of maintaining your positive electric charge and constantly on the look-out to pick up any loose electron that might float within range. But this is a relative world and you have no way of knowing whether you're moving or not. So, apparently at rest, you lie there cultivating your strong force, much as a poet might cultivate potential poems. Suddenly—in the far distance—a disorderly wall of jostling protons appears. It seems to be moving towards you at immense speed. Then Zap! you're through and out the other side. Unharmed and merely flushed with the excitement of a near miss. This is the experience of a proton in an experimental target. Even though there might be a million million protons in the beam and there were about 3×10^{19} (30 million million million) protons to the cubic millimetre in the target, the chances of you being hit were not high.

Again, think of yourself as a proton. You're quietly minding you're own business. This time we will tell you that you are travelling along with several other protons, in a beam of particles. You're being kicked electrically into a very steadily increasing velocity. (But as we said above you could know none of this if we hadn't told you.) Again, in the distance, an array of jostling protons appears. This time you're not so lucky. Zap! You collide with another proton. The resistance aspect of the strong force you are exuding is overcome, and the two of you fuse together in a loving embrace, your energy merging into one. But alas this state of cosmic excitement lasts only for 10^{-23} seconds of clock time. After that you divide again from your lover; but before doing so, in

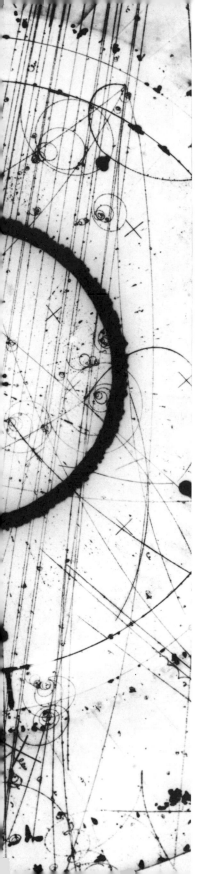

that instant of unity you revalue the energy of your embrace and, deciding that you are a little over excited, you restore the balance by throwing out one of your potential poems (a particle called a pion) and now you divide again from your lover and go on your way, slower, less energetic, but otherwise exactly the same identical quantum that you were when you started.

Here is the central fact of current high-energy physics. The break-through into the new world of quantum effects and relativity. The act of 'looking' at smaller and yet still smaller things has led to the situation where the act of looking creates new things. The energy of the act of looking transmutes into a new thing. This would seem to be a crucial turning point in the study of the physical world. The world stands revealed as process of a super-subtlety. The study of 'things' ceases, since as we will see 'looking' very hard merely produces an endless series of new 'things'. The whole strategy of the study of fundamental states shifts from the study of things to *the study of the ratios and proportions of the process itself*. This is why particle physics is concerned with interactions and events more than objects.

The fundamental situation at the sub-atomic level, which is to say the fundamental situation of everything everywhere, is that of a gigantic cosmic dance of creation and destruction, the ebb and flow of the 'ten thousand things' of the Tao te Ching. The Dance of Shiva of Indian philsophy. If we are bent on a description of it, our knowledge will be limited to discovering the choreography from which it takes its form—this ballet of creation and annihilation that arises from the no-thingness of the underlying quantum field.

Let's consider, through our word model, a more elaborate version of a part of the dance. This time you are a proton sitting in a bath of hydrogen. If you were a man-sized proton your nearest neighbour would be 300 miles away and your single electron would be whirling about solidly locked into an electric fence, which is in turn locked into the other electric fences of the other neighbour protons about 150 miles away. Suddenly, you see a particle coming towards you very very fast. ZOK! a very energetic incoming particle hits you and you merge together. You are moving fast but at less than the original speed of your fast friend. Together you are about twice as heavy as you were to start with, apparently much bigger, and also you are leaving a stream of bubbles in the liquid hydrogen as you go along. You may also notice that two other little particles fly off. Excited by the collision you hurtle through the hydrogen bath for what seems ages. A million million units of what is for you a natural unit of time pass before the first of your triple burdens of strangeness can be shed. A weak and slow untying of the embrace with which you cling to the extra energy of your frantic encounter proceeds. At last you yield and divide in two. A little pion veers away from you carrying your electric charge and leaving you feeling electrically neutral (and also, no longer leaving a stream of bubbles behind you as you go).

You travel on another million million units of your time (but only for a thousandth part of a millionth of a second of *our* time) and then the second of your strange bonds breaks and a neutral pion is born from your death as you are instantly reincarnated, still strange, but only once removed from your starting point and goal. You reach this in due time when the weakly felt but inexorable magic of change has run its course. You continue on your merry way until by disgorging a negative pion you regain your positive charge, rid yourself of your excess weight, lose all feeling of strangeness and emerge from this flurry of activity serene, calm, and a proton once more. As you sail silently into the blackness you can look back and observe all the various dancing partners and their dancing partners still briefly visible through the tracks of bubbles they and you have left.

But can we accept this description of what seems to be normal on the cosmic scale? Before we go on it may be beneficial to look in a little detail at the physical situation from which the descriptions of the collisions we just read were taken. All our information comes from collisions of one kind or another, and the key high-energy physics machines, after the accelerators which produce the beams of particles, are the detectors which display collisions in ways such that they can be recorded and studied.

One of the main detectors in use today is called a bubble chamber. As you can see from the picture its shape tells you nothing about its function. A bubble chamber such as this one at CERN in Geneva, consists basically of a huge tank full of liquid hydrogen contained between the poles of a very large electromagnet. The liquid hydrogen is convenient because it contains only protons and electrons. In this case the tank is two metres in diameter and

Above: the two-metre bubble-chamber at CERN

holds 200 gallons of liquified hydrogen.

The whole thing is like a pressure cooker in which, instead of a lid, a piston keeps the hydrogen at just below its boiling point. A stream of particles is sent into the hydrogen and then for a fraction of a second the pressure is lowered, and along the lines where the particles have disturbed the hydrogen atoms, the hydrogen boils. All in the same fraction of a second, cameras snap the lines of bubbles; the particles had gone long ago, but for a brief moment they have been made to leave behind them their own special kind of sky-writing.

But what then? The pictures are beautiful, but what do they reveal except that we live in a violent universe? How can something permanent and unchanging be found in pictures like these?

To begin to see what the pictures reveal means penetrating what they show more deeply than seems possible by just looking at them (physicists look at, and measure them, by the millions). For instance they occur in three dimensions. The pictures are records of an event, a happening; one picture can no more show the quality of the event than could one picture of a billiard game.

Below left, previous page, and on endpapers and title page: typical bubble-chamber pictures. Photographs of the stream of bubbles left by a variety of different particles following several collision events

The Particle Zoo

γ **photon:** a particle of light; electromagnetic radiation when it behaves as a particle; interacts with electric charge and electric currents, and with magnets; has one unit of spin.

ν **neutrino:** like the photon always travels at the speed of light; has only weak interactions; always spins (with $\frac{1}{2}$ unit of spin) about its direction of motion like a left-handed screw; has an antiparticle (which spins like a right-handed screw); because of its weak interaction is very penetrating, and therefore hard to detect; comes in two different varieties, one associated with electrons, one with muons.

e **electron:** the least massive charged particle; the number of electrons in an atom, and their configurations, determine chemical properties; configurations of electrons in solids are responsible for different properties of metals, semi-conductors, insulators, etc.; first identified in 1896; has one negative unit of electric charge; has an antiparticle (the positron) with opposite charge; interacts with electromagnetic field (i.e. with photons) and also participates in weak interactions.

μ **muon:** a sort of 'heavy electron'; similar in almost every respect to electron, but 200 times more massive; decays spontaneously with average lifetime of 10^{-6} sec into electron and two neutrinos (one of each kind), the surplus mass being released as energy of motion of the decay products; this decay is example of weak interactions; muons, electrons and neutrinos constitute family of leptons; all have spin of $\frac{1}{2}$ unit.

π **pion:** least massive of the hadrons, or particles with strong interactions; comes in three varieties with charges + 1, 0, −1 units; has no spin; principal carrier of nuclear force; decays spontaneously, the charged varieties in 10^{-8} sec via a weak interaction into a muon and a neutrino, the neutral variety via an electromagnetic interaction into two photons; pion mass has energy equivalent of 140 Mev (if electron$=\frac{1}{2}$ Mev).

K **kaon:** strange cousin to the pion; has no spin; has strong interactions; is massive enough (mass equivalent to 500 Mev) to decay into pions, but cannot do so via strong interactions because it has strangeness (and pions don't) so decays via weak interaction, with lifetime 10^{-8} sec (strong interactions never change the total amount of strangeness); there are positively charged and neutral kaons with one positive unit of strangeness, and their negative and neutral antiparticles have one negative unit of strangeness. These four kaons, the three pions, and a particle called the eta form a family of eight spin-less mesons.

p **proton:** stable constituent of atomic nucleus; positively charged; the number of protons in the nucleus determines its charge, hence the number of electrons in the atom, hence its chemistry, etc.; mass equivalent to 940 Mev; has spin of $\frac{1}{2}$ unit; has strong interactions and so can exchange pions (and other mesons) with other constituents of nucleus—this is the basis for exploration of nuclear force.

n **neutron:** neutral brother of the proton; also constituent of nuclei (except for hydrogen nucleus which is just a proton); protons and neutrons so similar except for charge that they are best considered as just different charge states of the same particle, the nucleon; number of nucleons in nucleus is called its atomic number (e.g. 14 for the carbon used to date archaeological samples, 235 for uranium in reactors); slightly more massive than proton, so that a free neutron can decay spontaneously—but only just, so it lives on average about 10 minutes before doing so. Is generally stable when bound in nucleus, but is still unstable in some nuclei which are radioactive (beta emitters); decay is via weak interaction into proton, electron and antineutrino.

Λ **lambda:** strange cousin of the nucleon; has no charge but like nucleon has $\frac{1}{2}$ unit spin and has similar strong interactions; however carries the same 'strangeness' attribute as the negative kaon; may be produced together with a positive or neutral kaon in strong interaction between nucleons; but having been produced can only decay via weak interactions, so can live as long as 10^{-10} sec; decays most often to nucleon plus pion; mass is equivalent to 1120 Mev; being more massive than nucleon is called a hyperon.

Σ **sigma:** strange cousin of nucleon; has three charge states (+ 1, 0, −1 unit of charge) and like the lambda has one negative unit of strangeness;

as with lambda may be produced in association with kaon by strong interaction, but decays by weak interaction to nucleon plus pion; mass equivalent to 1190 Mev.

Ξ **chi** (or **cascade**): cousin to nucleon; has two charge states (−1, 0); has two negative units of strangeness so cannot decay by strong interaction; and even via the weak interaction, which can change strangeness by one unit, cannot decay directly to nucleon plus pions (which are non-strange) so decays mainly to lambda plus pion; lives about 10^{-10} sec; mass equivalent to 1320 Mev; the nucleons (p, n), the lambda (Λ), the sigmas (Σ⁺, Σ⁰, Σ⁻), and the chis (Ξ⁰, Ξ⁻) are a family of eight particles which (like the family of eight mesons) can be regarded as different states of just one unit of baryon number, which is the particle physics name for atomic number; total baryon number of a system never changes, which is why the proton, as the least massive baryon, cannot decay.

Δ **delta:** excited non-strange baryon; produced in pion/nucleon collisions; is a state of nucleonic matter with internal energy different from the nucleon; has mass of 1230 Mev; comes in four charge states (+2, +1, 0, −1); is non-strange; has spin 3/2; can decay with strong interactions, so is very short-lived (about 10^{-23} sec) and is observed indirectly by studying its decay products; it is also called a resonance state.

Ω **omega:** spin 3/2 baryon with strangeness −3; has one charge state only, charge −1; together with four charge states of delta, three of an excited sigma, and two of an excited chi, makes a family of ten; its existence and properties predicted from relationships between members of such family, from properties of other members; has mass equivalent to 1670 Mev; lives about 10^{-10} sec before decaying via weak interaction, usually into chi plus pi; less than 50 have ever been seen!

ψ **psi:** discovered in 1974; unusual and unexpected in that, although massive (3100 Mev), and so with enough energy to choose between many possible decays, still lives a comparatively long time—at least 150 times longer than particles which have strong decays; has no charge and zero baryon number; has electromagnetic interactions; probably has spin of 1 unit; may be related to the families of mesons with spin 1 (one family of 8, one with just one member), in a scheme of relationship which adjoins a new quantity, called charm, to charge, strangeness and baryon number; has an excited state (ψ with mass 3700) which decays mainly into ψ and pions.

There are many other particles, often related in family groups of 8 or 10 members; some of the families are also themselves related one to another, differing essentially only in the amount of internal motion their members carry.

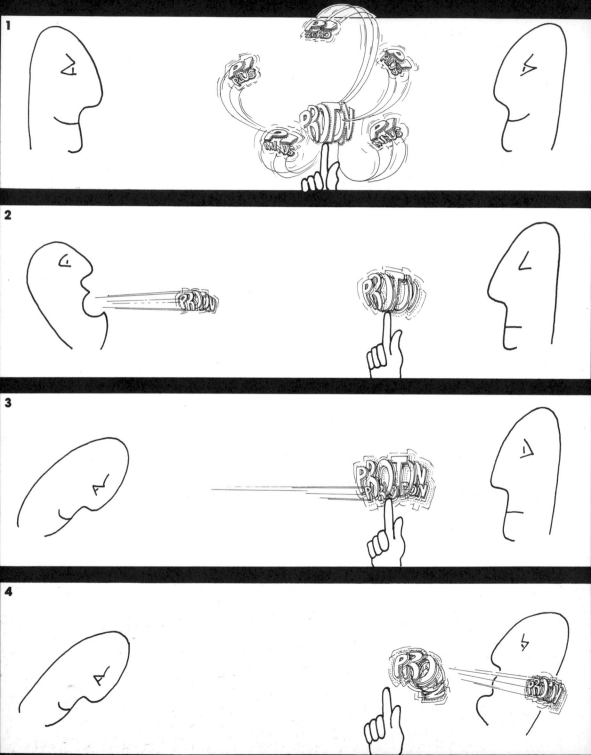

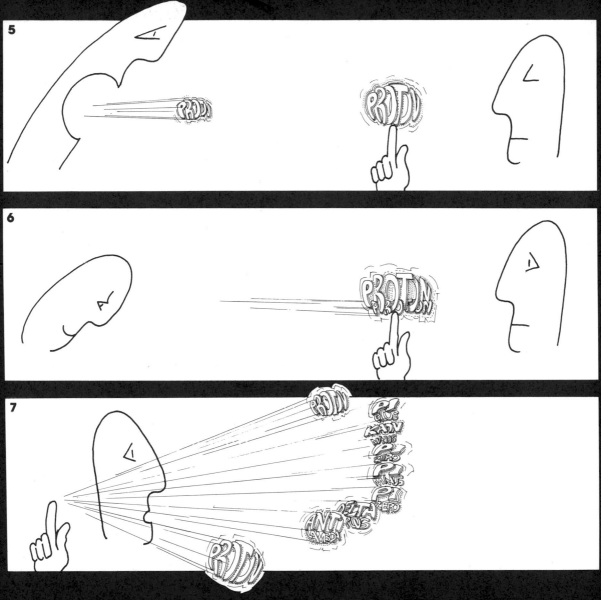

Fig 1 Virtual particles. The strong force manifest by the proton is often thought of as a cloud of virtual particles which carry the strong force

Figs 2–4 Low energy collision. A proton hits another proton in a relatively low energy collision. The protons bounce off one another unchanged

Figs 5–7 High energy collision. A very high energy proton approaches a target proton. After the collision, in addition to the original two colliding protons, seven more particles have been produced; some of the extra energy of the fast proton has turned into matter

Figs 8–10 The strong force that holds together a nucleus, here a deuteron (heavy hydrogen), may be pictured as virtual particles being exchanged by two nucleons either by the exchange of pi plus (Fig 8), or juggling with pi zeros (Fig 9), or the mutual exchange of pi zeros (Fig 10), or …

Fig 11 Similarly the gravitational interaction is believed to be mediated by the exchange of a particle called a graviton

Figs 12–15 The electromagnetic interaction is held to be mediated by the exchange of photons

Figs 16–19 The strong interaction. In this high energy collision, two protons collide.

After the collision one proton remains, the other has turned into a Delta plus (Fig 18). This rapidly decays into a proton and a pi zero (Fig 19)

Figs 20–3 The weak interaction, mediated by a yet undiscovered particle, the intermediate boson. After an average period of 10 minutes, a neutron … decays into a proton an electron and a neutrino

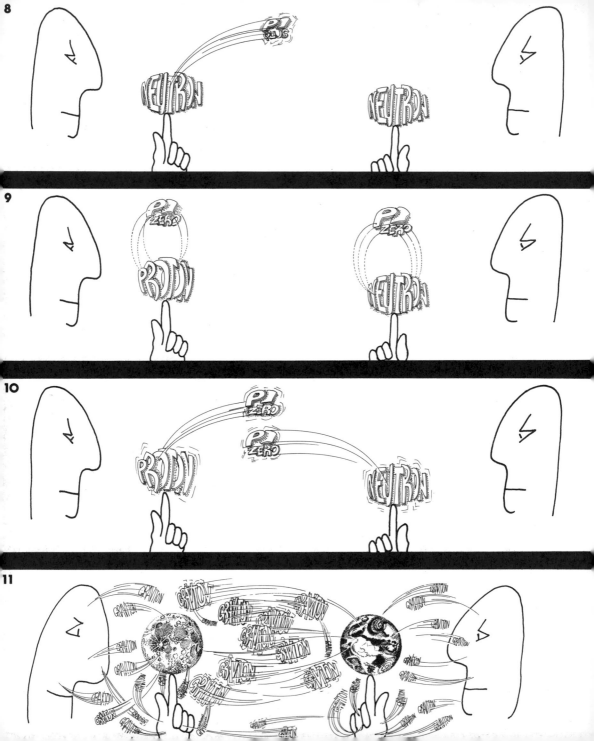

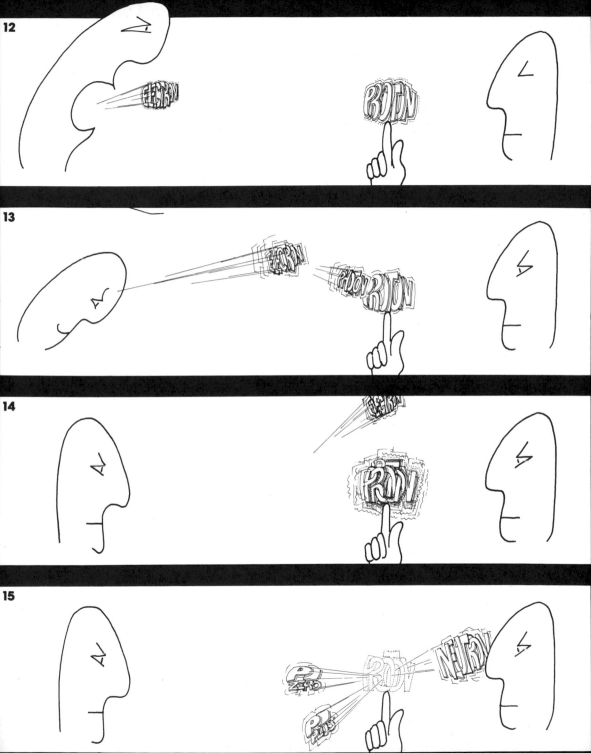

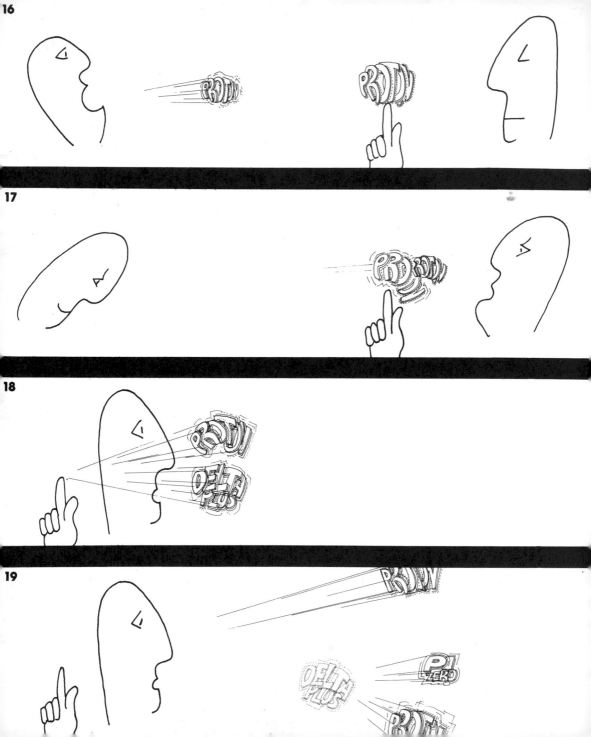

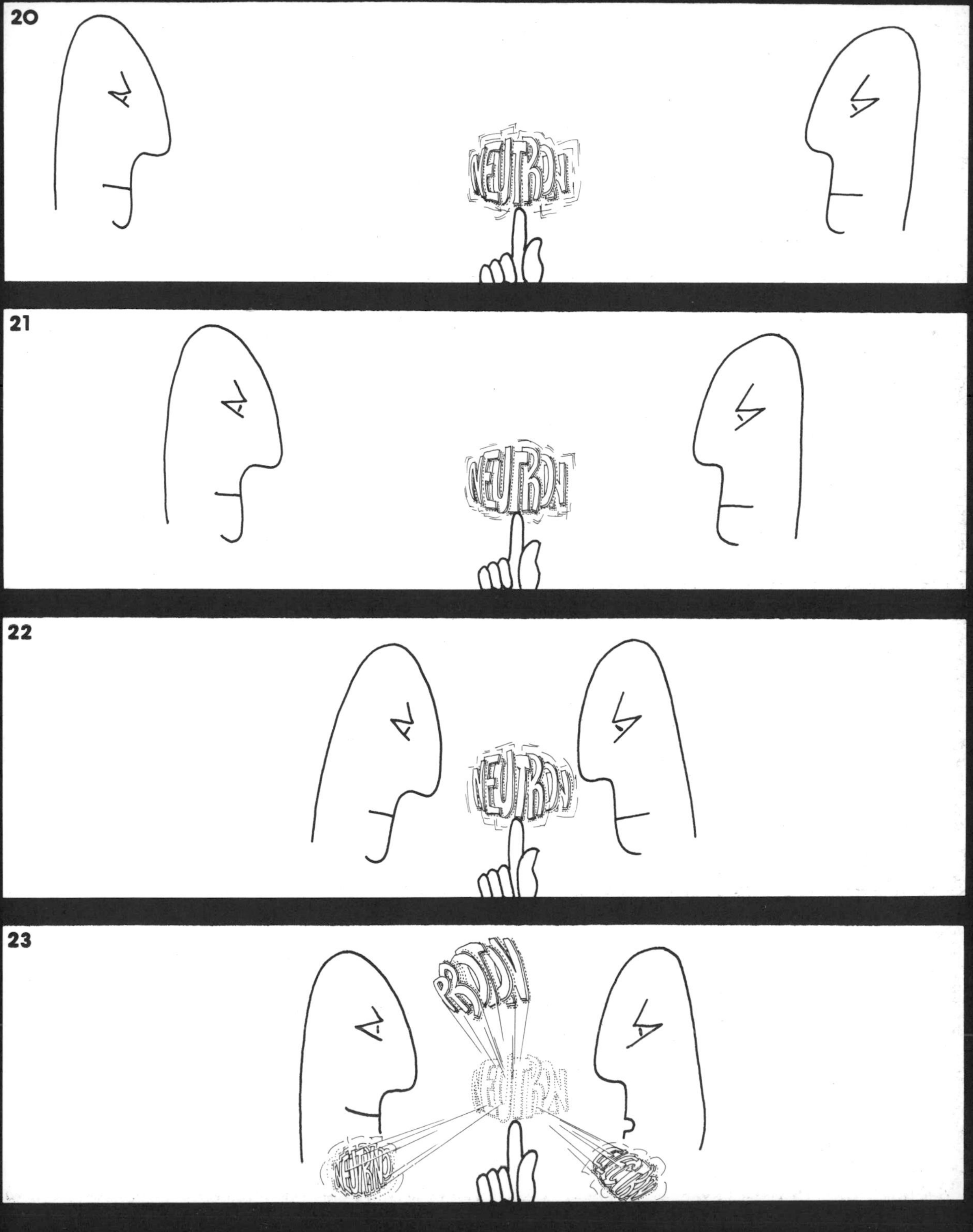
20
NEUTRON
21
NEUTRON
22
NEUTRON
23
PROTON
NEUTRON
NEUTRINO
ELECTRON

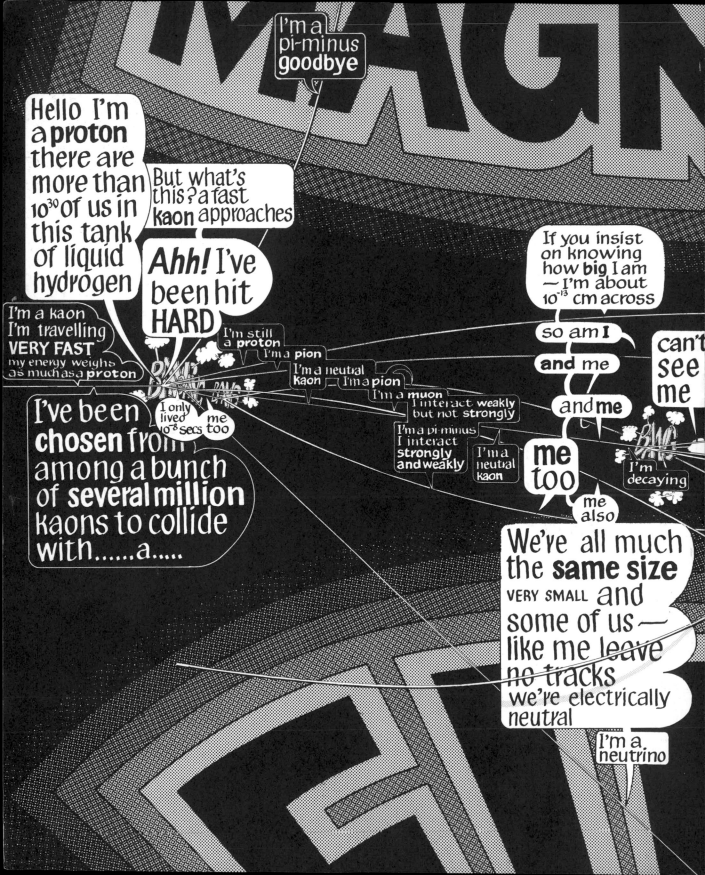

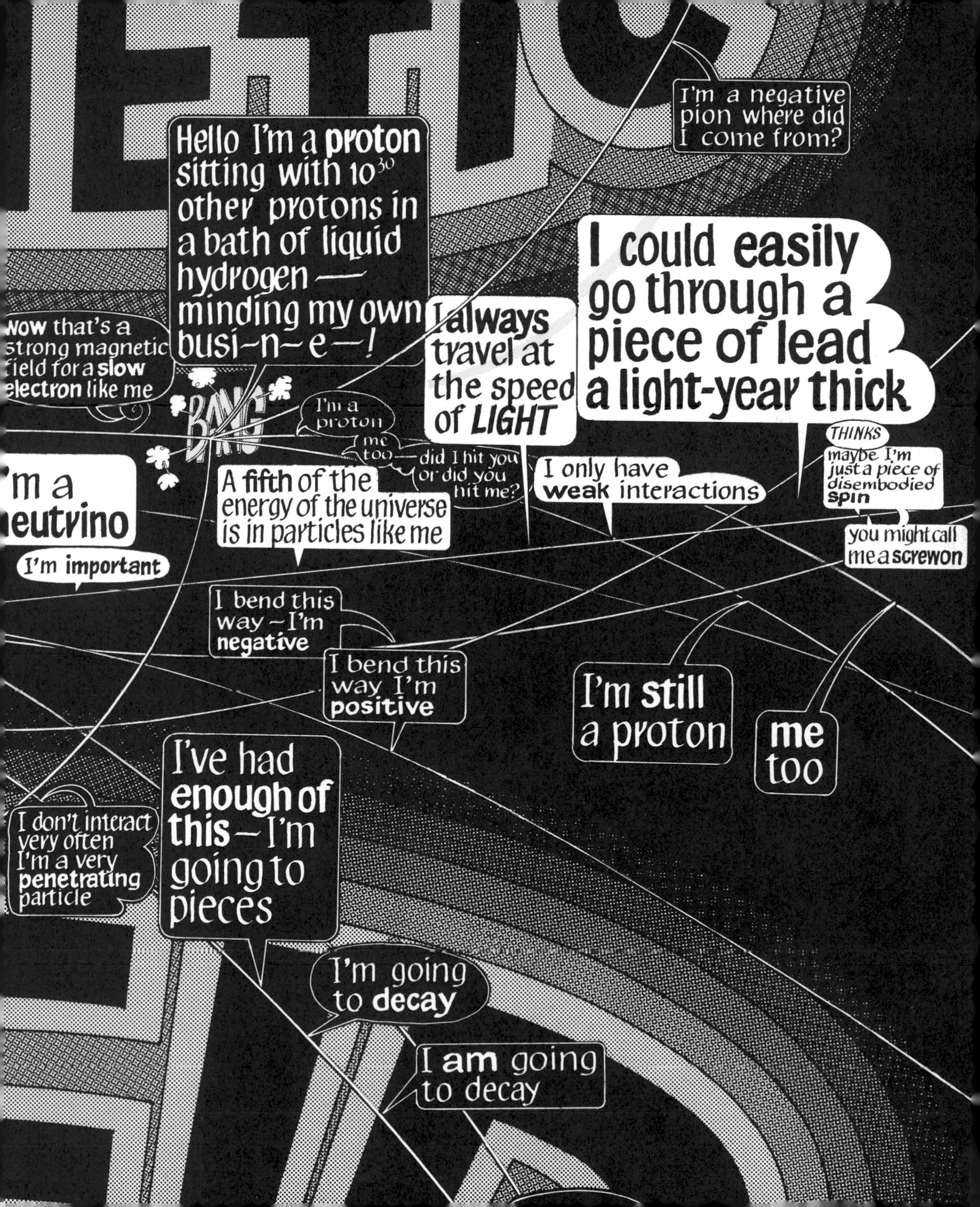
I'm a negative pion where did I come from?
Hello I'm a **proton** sitting with 10^{30} other protons in a bath of liquid hydrogen — minding my own busi-n-e-!
I could easily go through a piece of lead a light-year thick
I always travel at the speed of *LIGHT*
Now that's a strong magnetic field for a **slow electron** like me
BANG
I'm a proton
me too — did I hit you or did you hit me?
I'm a neutrino
I'm **important**
A **fifth** of the energy of the universe is in particles like me
I only have **weak** interactions
THINKS
maybe I'm just a piece of disembodied **spin**
you might call me a **screwon**
I bend this way — I'm **negative**
I bend this way I'm **positive**
I'm **still** a proton
me too
I've had **enough of this** — I'm going to pieces
I don't interact very often I'm a very **penetrating** particle
I'm going to **decay**
I **am** going to decay

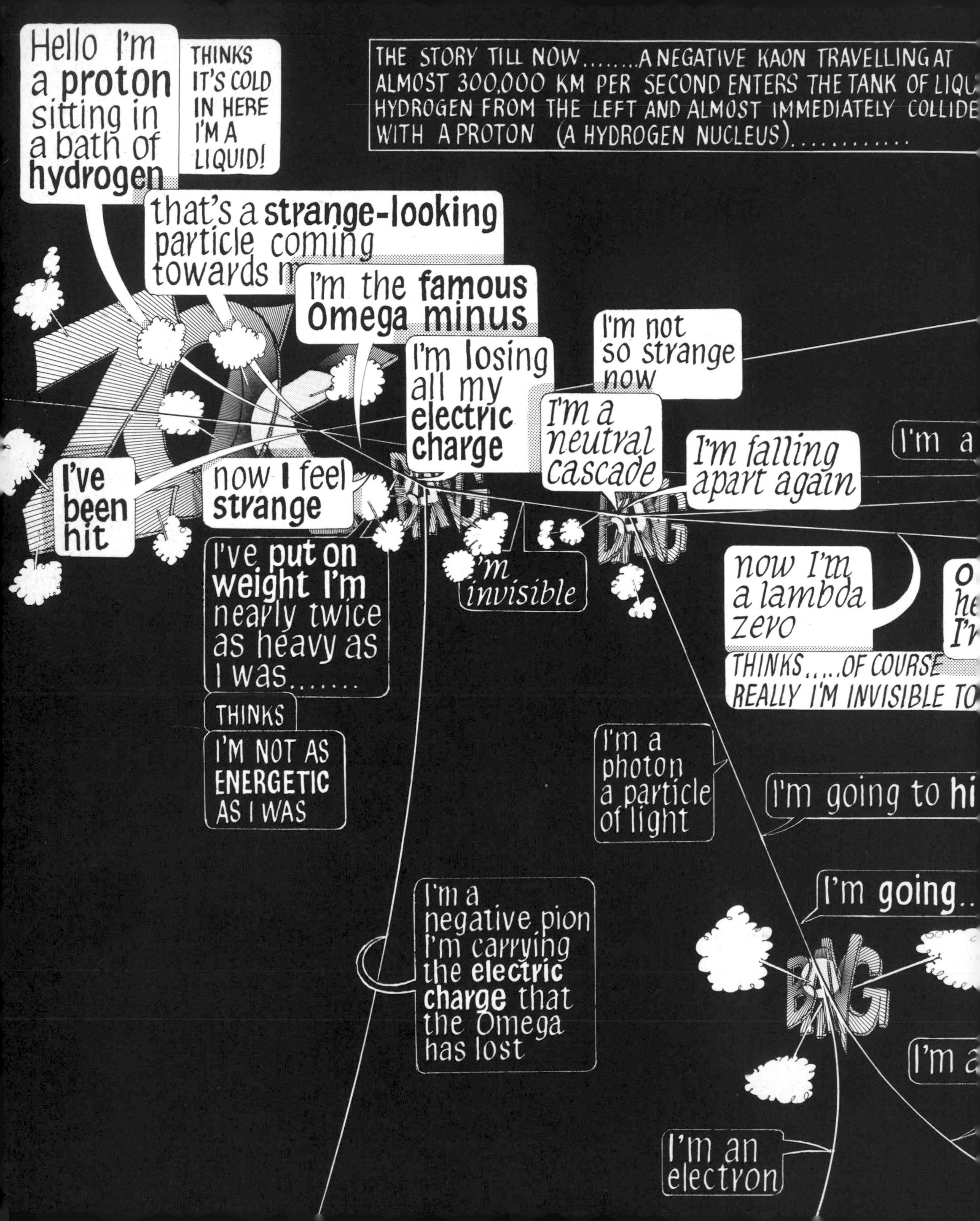

THE STORY TILL NOW........A NEGATIVE KAON TRAVELLING AT ALMOST 300,000 KM PER SECOND ENTERS THE TANK OF LIQU
HYDROGEN FROM THE LEFT AND ALMOST IMMEDIATELY COLLIDE
WITH A PROTON (A HYDROGEN NUCLEUS)............
Hello I'm a proton sitting in a bath of hydrogen
THINKS IT'S COLD IN HERE I'M A LIQUID!
that's a strange-looking particle coming towards m
I'm the famous Omega minus
I've been hit
now I feel strange
I've put on weight I'm nearly twice as heavy as I was.......
THINKS
I'M NOT AS ENERGETIC AS I WAS
I'm losing all my electric charge
I'm not so strange now
I'm a neutral cascade
I'm invisible
I'm falling apart again
I'm a
now I'm a lambda zero
THINKS.....OF COURSE REALLY I'M INVISIBLE TO
I'm a photon a particle of light
I'm going to hi
I'm going...
I'm a negative pion I'm carrying the electric charge that the Omega has lost
I'm a
I'm an electron

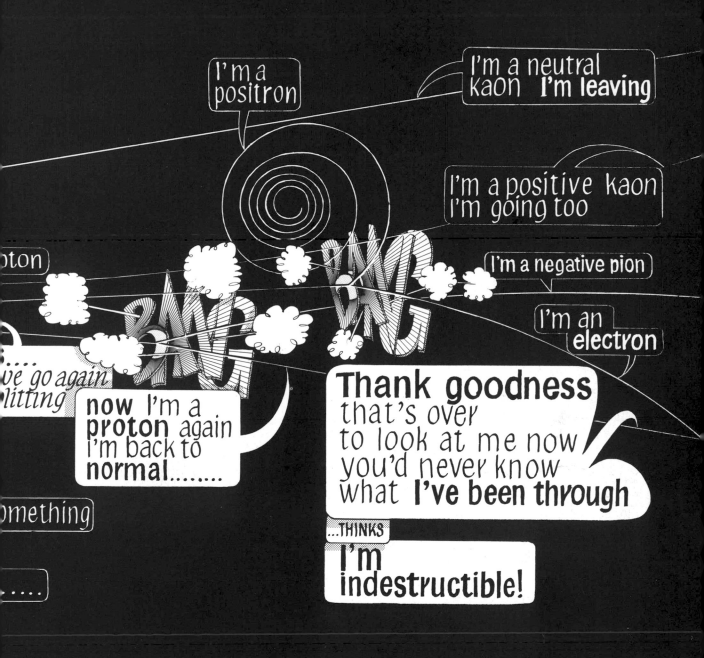

Starfield in the Milky Way

Most of what you have read up to now has been deducted from data contained in bubble chamber pictures such as those we've just seen, or from electrical data in a similar, but invisible computerized form. Literally millions and millions of similar pictures have been scanned and measured to provide the statistical basis for the ideas we've been discussing. What physics has been looking for amid the violence and apparent chaos is invariance, something that is constant and unchanging in the cosmic dance. What the physicist has found there is the subject of the next two chapters.

Have you found the idea of a cosmic dance useful? (It's not being suggested that there is a Choreographer in the wings; at the time of reporting no such Person has emerged!) That there *is* a cosmic dance of creation and destruction there can be no doubt. It is taking place at one level of energy or another all the time, everywhere. It's going on in the room now as photons pour in through the window or from the light bulb, and are absorbed and emitted by

the material they fall on.

In the sun itself, only one of unimaginable millions of similar suns, the dance is taking place with dazzling intensity. All the energy man has generated and consumed in the whole of history is no more than a match flame compared with even the minor scale of the sun's furnace. Here the performance never stops. It fuels the intense output of radiation that sustains life on earth. The protons, neutrons and electrons in the atomic ballet that is us and the world of matter, dance more sedately, but this solid state of matter is a local speciality, a one-nighter on the universal scale. On the cosmic scale it's the starquake level of energy transactions that is normal.

There is a cosmic dance. It's a dance unlike any man has known before. It has unique connections between the music and the dancers. In what other theatre does it happen that the louder and faster the orchestra plays, the more dancers are created?

143

9. Balancing the Books

The awareness that you may now be developing of the interior dynamics of the universe and especially the scale on which it operates is essential if one is to be able to hold in mind the relative scale of our human endeavours. The struggle of life occurs on the scale of a breeze filled with jostling pollen grains, each looking for a life sustaining place to rest, rather than as we tend to experience it, as a blind wrestling match.

But we are fortunate to be able to apprehend imaginatively the extent of the world in which we are immersed; this in itself can bring nourishment, but not necessarily (as we will see in the final chapters) a completely balanced diet. The penetration by physics to the level of knowledge about the physical universe that we have been looking at is a major creation of intelligence. It is a feat of explanation which has given decisive shape to our lives.

But this mountain of data, experiment and theory that is physics and in whose shadow we can hardly avoid living our lives, has buried within it yet another level of subtle abstraction. Did you think the subtlety had come to an end? In fact this is where the story really begins, because now we are beginning to reach the physics of the nineteen fifties and sixties. It's difficult, but remember it's difficult for physicists too; so relax, keep an open mind.

There is no mystery in what we are moving on to. We began by noticing the separateness around us, and moved on through increasingly simple explanations for the structure of the world until we discovered eventually that thinking of structure as a composition of 'things' was meaningless. We have now come to the point of appreciating the true meaning of the word nothingness. But nothingness here does not mean emptiness. What nothingness means here is the behaviour of space itself, a dynamic process of vibrating energy states. And the essential nature of this process (which, remember, *is* us), is its total interconnectedness, interdependence and interpenetration.

That on the face of it might seem to be that. Everything's all connected, goodnight! If knowledge of the connectedness was all that mattered then we could leave off now; but if, as seems obvious, we want to begin to *realize* what this connectedness means, then the deeper subtleties have much to teach us.

We can begin by returning again to the ways that energy can be manifest. Energy can appear in many forms: the energy of motion—the kinetic energy of the motion of a whole body, or the internal motion of its constituents, i.e. heat; the energy inherent in mass; and the potential energy wound up in a spring, or held in a magnetic field.

In interactions, reactions and other processes, energy can freely convert from one form to another *but* the total amount of energy never changes. Ever.

This overriding rule which says that the total energy remains constant or is 'conserved' is one of several such rules which science has identified in nature, and they are of course key insights since they inform and shape everything. Is this clear? Take your time. It's not too difficult to get an intuitive sense of how energy may be required by nature to 'balance its books'. This is important for two reasons: one is that it allows observations that predict the outcome of a reaction if the ingoing situation is known, and secondly it means that certain states of affairs in reactions or processes are forbidden.

For instance if, in a simple reaction, the total energy measured before the reaction was more than was observed afterwards, then this clearly means something is wrong somewhere. Suppose that two particles collide together. Their total energy is known. But after the collision when their energy is measured, it's *less* than it was before the collision. Where did the extra energy go? It must be there *somewhere*. A further observation might show that a previously unsuspected particle had been created at the point of the collision. It is then an easy matter to *infer* from the total energy sum, what energy the missing particle must have. In this way a combination of measurements of the total energy in a collision can lead to the discovery of new particles. Just this process has led physics to lay bare a whole zoo of new fundamental particles; the list at the moment totals about 300, all of which have been produced and detected in the kind of collision situation we have been looking at.

The production of new particles in such numbers has of course caused a hiatus in physics. What does it mean? Why are there so many? Why are they so similar? Some of the particles are so unstable and have such short life-times that their existence is manifest only as a resonance between the fragments into which they decay. Others are a little more stable and seem more tangible, more real. This is where we find the frontier of physics. No one really knows why there should be so many particles but there is an intense preoccupation with discovering a simple explanation for the nuclear zoo. Yet again, a new extension of the senses, this time through higher energy particle beams and improved detectors, has extended the apparent diversity of the world. And the search for simplicity continues.

We've seen earlier how there is a certain threshold in collisions between particles. Below the threshold the two colliding particles meet, then go their separate ways, unchanged. The balance in energy has been met by the slowing down of one particle and the speeding up of the stationary one.

Above this energy threshold, at very high energies, everything changes. The energy is so great that an alternative way to balance the energy becomes possible. Some of the energy of the collision can be turned into the mass-energy of a new particle in the act of collision. A new particle is created from the quantum field, containing just the correct amount of energy to balance the total. The strange thing is that though the collisions can be made to occur with a smoothly rising gradient of energies, particles are produced only here and there on this energy gradient. It is as if we were using a torch with a brightness control, to look at the contents of a shed. At dim brightness, we find apples;

then as we turn up the torch a little we see nothing; then at a certain increase in brightness we see oranges but a little brighter or darker nothing, and then as we increase the brightness further at certain quite well defined intervals we see a succession of (heavier) fruit; until after quite a long period of increasing the brightness to a very high intensity we are 'seeing' pumpkins in the shed! This may seem a light-hearted way of looking at the high energy physics process, but if we were to add numbers and labels it would be a description with which physicists are familiar.

Another useful way of approaching this production of particles is to think of it entirely in terms of excitation. If electrical energy is forced into sodium vapour past the point at which the sodium atoms can accept it, then the atoms emit particles, photons, of the characteristic orange frequency and colour often found in street lighting. This is true for all atoms; each has a characteristic 'colour' or 'colours' that it emits when it becomes overexcited. Can you see how similar this is to what is happening in proton collisions? The proton becomes over-excited and, at certain levels of excitation, emits particles only of certain definite 'colours'. The excitement doesn't last long for many of these particles; they usually decay quite quickly into less energetic states.

This situation that the physicist faces is a complex one. The large number of particles that have recently emerged from high-energy collisions gives only a hint of the multitude of differentiations that the new accelerators and detectors have revealed at a fundamental level in nature. But as we've seen this is only one phase in the process of understanding. A new machine is a new window. It extends the range of what can be seen. We must not forget that alongside the new window come further developments of the extensions of mind: thinking and memory. In fact, superfast computer logic and memory is an integral part *of* the new window. The machine's arrival is also paralleled by the further growth of physics theory. New extensions of mind and thinking continue to emerge in the form of new mathematics. It's perhaps only through mathematics that the new situation can be comprehended. But can it be comprehended? Is there simplicity within the newly discovered diversity? If it is there, how can it be revealed?

Think of yourself as a physicist and look at the set of images on the next two pages. They too have a certain orderliness, but abound in complexity. Think of them as the raw data which has emerged from many particle physics experiments all over the world.

While at one level we might be disposed to accept the images for what they are, and look on them with what an American psychologist has described as 'unconditional positive regard' (a quality not unlike love!), curiosity may lead us to search for something more fundamentally simple, a base pattern below the surface patterns.

Right and overleaf: images made by children using designs from the Altair series of pattern fields

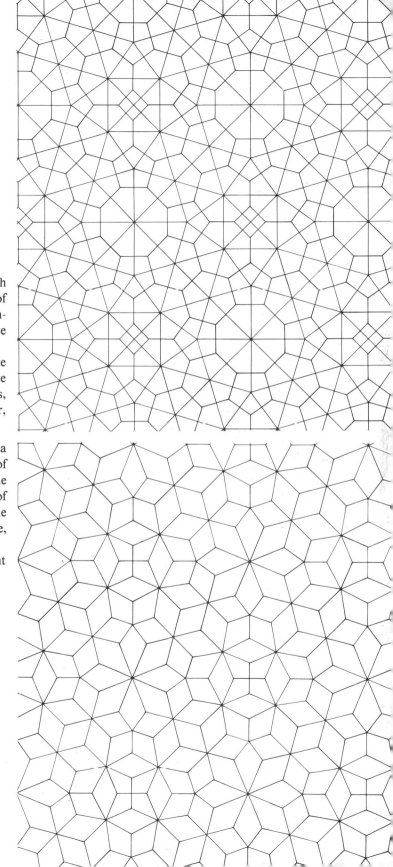

We might try an experiment: use this search for a 'base pattern' as a model of the kind of process by which particle symmetries may be uncovered. (Though remember that here we are dealing only with simple geometry.)

Can you see what the base pattern of these images is? Can you see into the patterns? There are many triangles, but also squares, pentagons, hexagons, heptagons and octagons and other, semi-regular figures (there's a clue here).

To pursue our analogy: the first thing that a physicist does is to compare the different sets of data to see if there are any correspondences. One way of doing this is to superimpose one set of similar images on another; and indeed, if the images given here are enlarged to the same scale, they can be superimposed.

Over the page, the two patterns on the right have been superimposed. What can you see?

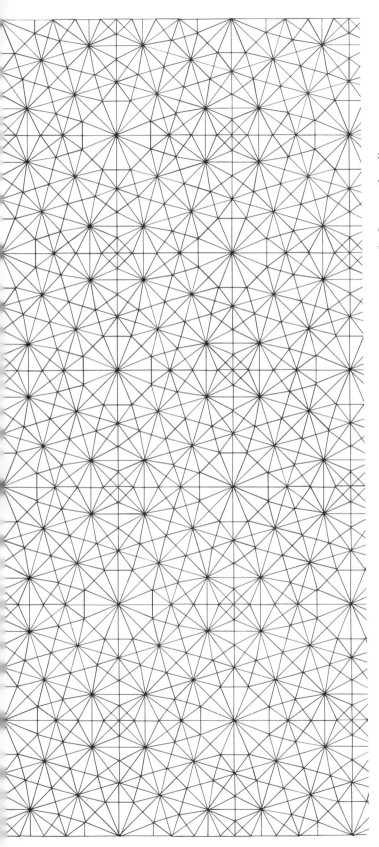

This superimposition suggests that there may be some underlying array of circles. Indeed, it turns out that all of the patterns are generated from the same linked set of circles, of only five sizes.

Finding the circles was the big jump, the first order of simplicity. Can you see what the next order is? Try to pencil it on.

By successive reductions (see below), we find base patterns of, first, one square array contained within four large circles, then one-quarter of this array, and then half of that, leaving a cell of five circles with their centres lying on the sides and points of a right-angled isosceles triangle. The relationships of the circles are unique and appear to have first arisen in the decoration which is characteristic of Islamic art.

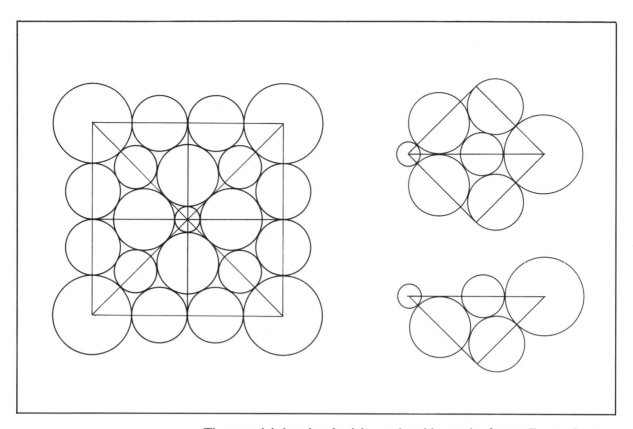

The material that the physicist works with now is *change*. Events. Interactions. Processes. But where does the possibility of comprehension of this change lie? Can the continuous apparently chaotic changes that elementary particles display ever be comprehended? Can something be found within the flux of change that is unchanging? These are the questions that particle physics is trying to answer now.

The tool that the mathematician uses to study invariance (or, non-change in the midst of change) is the theory of symmetry. The first kinds of symmetry to be brought to bear on the complexity of elementary particle behaviour are those that show that during a physical process some particular quantity *remains constant;* i.e. its total value remains unchanged after the process from the value it had before the process. As we saw earlier in the chapter, the quantity is, so to speak, *conserved*.

As a preparation for the next chapter on the more complex symmetries, which calls for a certain amount of mental gymnastics, we should first exercise ourselves a bit. Let's take a look now at those factors which were first seen to be constant and unchanging in particle interactions.

In collisions, as we've seen, the total energy that is involved remains constant no matter what goes on in the collision. There are several other quantities that also obey this kind of rule, which are somewhat different. Total momentum for instance remains constant. (By momentum we mean 'quantity of motion'—the momentum of, or 'quantity of motion' of, a polystyrene foam

While the principal of conserved properties of particles involved in collisions is not difficult to appreciate in a general way, a more complete grasp of what is involved means looking in detail at one collision and adding up the numbers that are conserved.

In the collision above, at (1), a 14 Gev kaon hits a stationary proton in a hydrogen bubble chamber. After the collision we find a positive kaon, two neutral kaons (one of them an anti-neutral kaon), a positive pion, a negative pion, and a proton. This is more conveniently written out by physicists as follows: $K^+ + p \rightarrow K^+ + K^0 + \overline{K}^0 + \pi^- + \pi^+ + p$.

This is a *strong interaction*.

Let's look first at the *electric charge*:

before the collision	and after the collision
$K^+ = +1$	$K^+ = +1$
$p = +1$	$K^0 = 0$
	$\overline{K}^0 = 0$
	$\pi^- = -1$
	$\pi^+ = +1$
	$p = +1$
Total: +2 electric charge	Total: +2 electric charge

Electric charge adds up to the same before and after the collision, i.e. electric charge is conserved.

Now we look at *Baryon* number:

before the collision	after the collision
$K^+ = 0$	$K^+ = 0$
$p = +1$	$K^0 = 0$
	$\overline{K}^0 = 0$
	$\pi^- = 0$
	$\pi^+ = 0$
	$p = +1$
+1	+1

Baryon number also adds up to being the same before and after the collision.

Now we look at *hypercharge*:

before the collision	after the collision
$K^+ = +1$	$K^+ = +1$
$p = +1$	$K^0 = +1$
	$\overline{K}^0 = -1$
	$\pi^- = 0$
	$\pi^+ = 0$
	$p = +1$
+2	+2

Hypercharge is also conserved.

Let's do the same with the *weak interaction* occurring at point (2) in the picture. Here a positive kaon decays into a positive muon and a neutrino, which is usually written as $K^+ \rightarrow \mu^+ + \nu$.

Charge:

before the decay	after the decay
$K^+ = +1$	$\mu^+ = +1$
	$\nu = 0$
+1	+1

Charge is conserved.

Baryon number:

before the decay	after the decay
$K^+ = 0$	$\mu^+ = 0$
	$\nu = 0$
0	0

Baryon number is conserved.

Hypercharge:

before the decay	after the decay
$K^+ = +1$	$\mu^+ = 0$
	$\nu = 0$
+1	0

Hypercharge is *not* conserved.

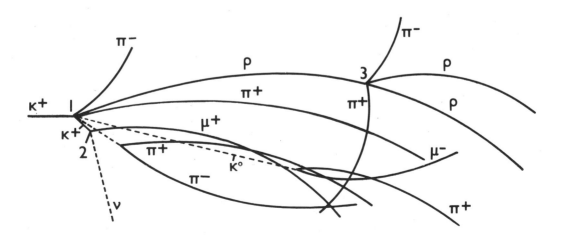

Left: an example of conserved properties in a particle collision event (shown in diagram). Some properties, like charge and baryon number, are absolutely and universally conserved by all interactions, strong, electromagnetic and weak. Others, though conserved by strong interactions, are violated by weak interactions—like hypercharge. But this violation is itself orderly and reveals more of the pattern of relationships between the particles and the properties they bear

Above: a photograph of the bubble-chamber event from which the drawing on p. 152 was derived. The original lines of bubbles were 2m long!

brick being shot out of a gun is obviously greater than that of a real brick gently lobbed.) So the total momentum—mass times velocity—in a collision doesn't vary, it remains constant. This is already a subtle quality to grasp; the others are even more abstract. Relax, take it slowly. Don't run away.

No one knows what 'electric charge' is. This may seem surprising, but it's true. What is known is that it, too, is conserved. If a positively charged and negatively charged particle go into a collision, their total charge, i.e. plus and minus added together, is zero. Then, whatever the outcome of the collision, the total electrical charges of the particles coming out will also always add up to zero. This particular conservation rule is made much simpler by the fact that, up to now at least, most particles have single units of charge: either none (0), or negative (−) or positive (+) units. The important point is that all charge is in integer multiples (and for fundamental particles *small* integers −1, 0, +1, +2) of a basic unit of charge—which seems to be an absolute constant throughout the universe and throughout all time.

Yet another quantity which is conserved, so far as is known, is labelled *baryon number.* Baryon number is the physicists name for what chemists call atomic weight. It's an expression of the size of a particle and is also commonly labelled, like electric charge, as positive, negative or zero.

There is also another quantity, *hypercharge,* which, like charge and baryon number, comes in simple integer units. But unlike them it is not absolutely conserved: it can change in weak interactions. Hypercharge conservation explains why certain particles are always accompanied by certain other particles in strong interactions.

Hypercharge? Electric Charge? Baryon number? This may seem somewhat resistable information, but we shouldn't fear it. It seems obscure because it's physics shorthand (and only part of the list, too; there could be added Lepton Number, Isospin, U-spin, Charm, Strangeness, and so on). Accept it as shorthand. What it *is* is less important than what it *means.* So let's ask ourselves—what does it mean? It seems to mean that certain quantities in collisions are conserved. But what exactly does 'conserved' mean? This is the crux. And it is difficult too, because it's possible to read about conserved quantities over and over again without it coming clear. *Something that is conserved is something that is unvarying* (or, non-change, when change is going on). Hypercharge, baryon number and charge are constant qualities that can be observed to remain the same whatever the depth of the violence of the events in which they are observed.

Here's a simple example of invariance in the midst of change.

Take a glass of water and a glass of milk. Using a spoon, take one spoonful of water and put it in the milk. Now, take one spoonful of milk and put it in the water. Now, repeat this operation ten or twenty times, but always once for each glass. Now, take it slowly—the amount of water in the milk glass . . . *minus* the amount of milk in the water glass . . . is ALWAYS ZERO, i.e. it is conserved. The equation for this if you like is M−W=W−M=0.

Now here we can see there are two things going on. One: we are *measuring* the result of a series of *operations.* Two: the *measurement* reveals that there is a symmetry of *movements.* The *operations* of transferring the milk from right

to left and left to right are *symmetrical*. Left and right movements are balanced.

The important point to note is that the conserved amount is independent of a whole range of superficial movements that might be made during the transfers from one glass to the other. For instance, the symmetry of the milk and water operations is independent of how long it takes for the transfers to be made, how far the spoon has to travel, how hot it is, whether in making the transfer the spoonfuls describe some fancy gyrations, or whether the water was stirred. All that matters is that the unvarying amount of water in milk and milk in water *is* conserved. The unvarying, conserved quantities charge, hypercharge, baryon number, strangeness, etc. have likewise been arrived at by simplifying the behaviour of elementary particles through considering them as operations.

Has it gone in? No? It's simpler than you think. Read it again!

Why should we be taking this amount of trouble over some apparently rather obscure mathematics? Why are physicists so preoccupied with conservation laws?

The answer is simple in a way. When physicists embarked on their search for a fundamental constituent of matter, they started looking for what we are *made* of. They found, as we've seen, that we are made of energy. That what there is, is energy. All else on the cosmic scale is transient. It is appearance. But within this appearance that on the universal scale is so transient and subject to total change, there are a few islands of permanence. The conserved quantities. On the cosmic scale *they are the only things that are permanent and unchanging*.

Opposite: family tree of elementary particles. All particles are believed to have anti-particles with equal but opposite properties. It should be remembered however, that they do *not* have negative mass. And some particles are the same as their anti-particles

154

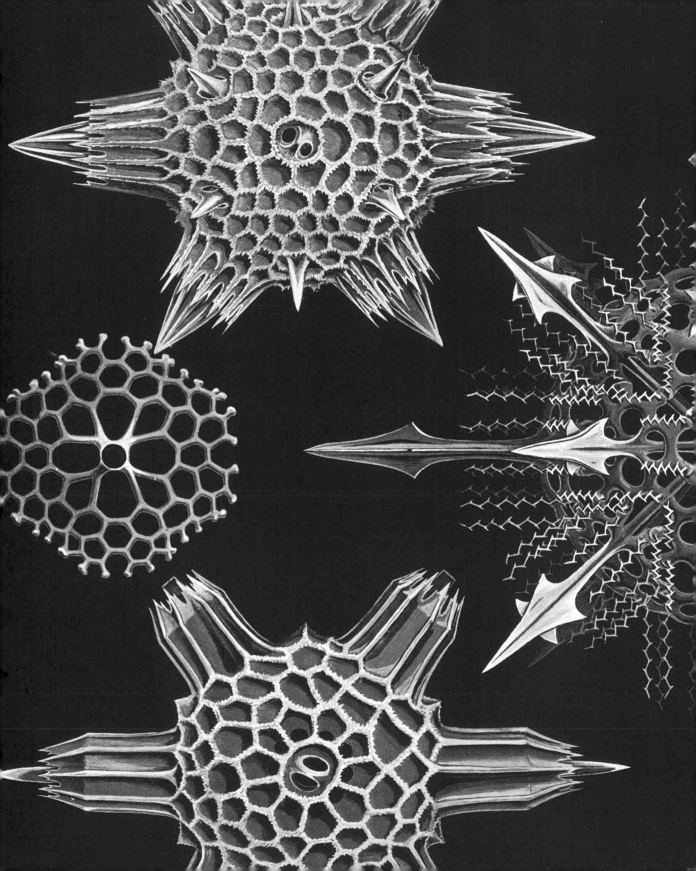

The chair under your body supports your weight. The printed page you are reading supports these words. But, as we have come to see, chair, book, body and words are no more than different flavours in a cosmic soup of particles – though by now even their solidarity may have receded a little!

Yet even this view of the world as a confection of tiny particles distributed through space, now turns out to be inexact (if useful). As we've seen, the nuclear 'zoo' of elusive and extremely short-lived particles, and the dance of creation and destruction in which they are manifest, points to an even more intangible universe, one of probabilities and resonances, rather than objects and things, however tiny. Above all it suggests that, though at one level the universe seems to be a collection of fundamental 'things', building blocks of matter, at root it is really a dynamical process. Of course to suggest this is not to deny the existence of atoms or particles, it's just that we are approaching yet another shift of emphasis from surface appearance to what lies within. From diversity to simplicity!

What do we mean by dynamical process? What informs particle behaviour? We've seen that there are things in it that don't change. What is it that is *ultimately* unvarying in it?

Consider the particle interactions as a game of snooker being played between two good players (and we assume of course that we have no previous knowledge of the game). If we look at what occurs in a relatively gross way, we might conclude that all that happens is that a lot of coloured balls are there on a table, and sometimes they move and hit each other; from time to time one disappears, and after they have all but one disappeared, they all reappear again. As our understanding of the whole game deepens, we begin to see that there are a variety of rules. The balls can have number and colour values assigned to them; the players alternate with each other; there are preferred strategies, and so on. It's this overall view of the whole game of elementary particle interactions that we are speaking of as a dynamical process.

Dynamical processes are characterized by order and symmetry, and it is to symmetry that the physicist increasingly turns as he seeks to understand the diversity he faces. It is here that the break with the past can be most clearly seen: much of the physics of the past (and much of everyday life for most people today) is centred on *objects,* on describing and controlling things. But modern physics is more concerned with *relationships* and *pattern.* (So also is modern mathematics—whereas the old maths dealt with circles, the new maths is interested in roundness.)

In an earlier chapter we saw that the events in which elementary particles participate are ruled by conservation laws, and that through these laws a preliminary classification of particles could be made. But now, with so many new particles turning up, a broader and deeper type of abstraction is needed. To return again to our billiard table—particles analogy, 'rules' might help us describe the game, but to understand the nature of the game the deeper simplicity of symmetries is needed.

In physics this search for symmetries takes the form of labelling particles according to the behaviour they regularly exhibit. In this way, certain behaviour that is unvarying, no matter what occurs in an interaction, has been

Previous pages: illustration from Die Künstformen der Natur by Ernst Haeckel shows extreme symmetry of various species of Radiolaria, a type of marine Protozoa

discovered—and labelled charge, hypercharge, baryon number or strangeness. Groupings of these labels can be studied, and this leads to a search for a super-order among the equations which express this invariance of behaviour. The results of this search for super-order then emerge as symmetries. A physicist would say that symmetries are a super-order of groups of transformations, which in turn refer to properties of the process itself (as it's observed). This is difficult to absorb. Think of it as a kernel within a kernel within a kernel situation. Give it time to unfold.

What do we mean by symmetry? The whole idea of symmetry is more than likely unfamiliar, so before we come to look for these kind of relationships in the properties of sub-atomic particles it may be useful to pause and consider just what we mean by symmetry. It may help to work our way through the possibly unusual ground of another kind of symmetry to which we all respond very directly. We will take it step by step. For the full benefit don't jump to the end.

To gain some insight into what symmetry *is*, we may usefully begin by having some contact with *proportion*, through an old idea that's still very much with us, as you'll see.

We'll begin by making some playful manipulations of simple numerical relationships. For instance, the series:

is linked in *geometric proportion*. Each number is two times the preceding one. In geometric proportion each successive term is multiplied by the first term.

Is there another series of proportional relations which we could superimpose on or add to these?

It turns out there is, and it's called *arithmetic proportion*. This is only a little more tricky. The arithmetic mean is the number which is halfway between the two extremes (the average). Now we can add a second series:

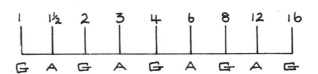

Now, deep breath (it's worth it in the end), there is another kind of proportion which is compatible, i.e. can be added into the two which we already have here. It's called *harmonic proportion*.

To find the harmonic mean between two terms (e.g. 2 and 6), we multiply together the two numbers (2×6=12) and then divide by their average (12/4=3). This gives us a term which exceeds the one before, and is less than the one following, by the same fractional proportion—for example, in the series 2:3:6, 3 exceeds 2 by *half* of 2 and is less than 6 by *half* of 6.

If we take the harmonic mean of our original series *(Series 1)*, we get:

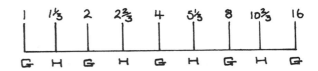

If we continue in our playfulness we can look again at *Series 2:*

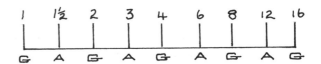

159

and derive the arithmetic mean of the geometric-mean-plus-the-arithmetic-mean:

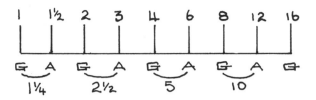

and from this new series a further arithmetic mean can be derived:

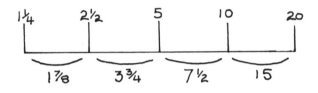

Also, if we go back to *Series 3*

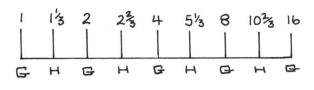

and derive the arithmetic mean of the harmonic-and-geometric relationship we get:

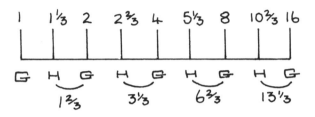

We could go on and on indefinitely, dividing and subdividing the series still further, and yet still maintaining total interrelationship between each term and all the other terms. But it may be apposite to stop here and put together all the numbers we have produced in all six series:

What we now have is a long list of numbers linked together in harmonic, arithmetic and geometric proportion. Can you see anything familiar about them? Almost certainly not. But it would be entirely another matter if you were to 'hear' the numbers as the *proportion* defining the vibrations of the strings on a musical instrument. You would then instantly recognize them as four octaves of the musical scale (the scale of 'Just Intonation').

Finally, if you are not altogether convinced, we can turn these *proportions* into string *lengths,* by turning them into fractions of the length of a single string (thus defining the fret positions on, say, a guitar).

Here are the first eight *proportions:*

they can be converted to *lengths* by making them fractions of the string length, i.e.:

proportions

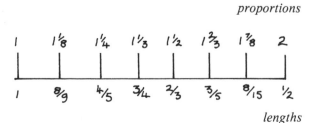

lengths

160

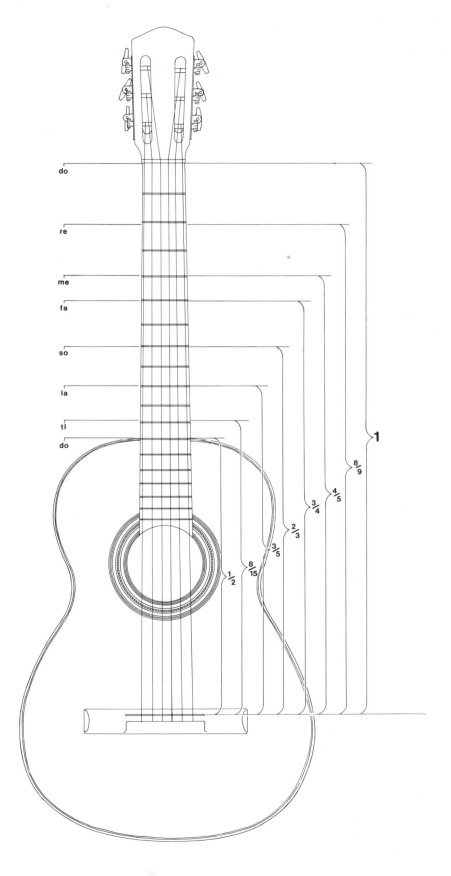

And finally they would look like this on a guitar.

If we were to play these string lengths exactly as they are defined here, we would immediately recognize them as the notes of the Western musical scale*.

Hopefully, it should now be clear that it's these multiple relationships of the musical scale that we understand intellectually as a symmetry and that we also directly hear and experience as musical symmetry. Any combination of musical notes enjoys this multiplicity of harmonious internal relationships, though some of them are more obviously consonant than others.

When we speak of symmetries in this chapter, it's to this kind of web of interrelationships that we refer. However, to provide a truer analogy for the kinds of symmetries that refer to particle states, we ought to include not just these proportional relations of pitch, but harmony and composition. The system of harmony given above is very old and was known in ancient Greece. Plato used similar ideas to convey his sense of the structure of the world, expressed as vibration and proportion. The symmetries of particle states are perhaps a modern, correspondingly subtle, expression of the same idea.

---

*On fixed-note instruments—guitar, piano, etc.—or in orchestral playing, the equal or well-tempered scale is used. This is based on 12 exactly equal semi-tones, each in the ratio of a very complex fraction. This scale, a compromise to which we have become accustomed, is out of tune with natural vibrations, but enables modulation.

At first sight it might seem that symmetries in particle physics are just a crystallization of the lower orders of simplicity of conservation laws; but this is misleading. It's true that the more abstract notion of symmetry absorbs and deepens some of these concepts of constancy and invariance. (For instance, the symmetry associated with the invarying quality of charge is a matter of 'one more' or 'one less': -2, -1, 0, 1, 2, 3, etc., whereas the symmetry of parity or of anti-matter/matter is simply one of opposition: 'this-is-the-opposite-of-that'.) But another kind of symmetry with which physicists concern themselves is more like that possessed by a sphere. Stay with it! Just read. Don't try to understand. Absorb it.

What relationship do the points on a sphere have? Each point of the sphere is evidently like each other point—the points of a sphere are all equivalent to one another. And the points of a sphere are continuously connected one to another. By a rotation of the sphere, any point can be carried smoothly to the position previously occupied by any other point. The sphere can be rotated

and yet remains the same sphere. The rotations can be generated first about one axis (arbitrarily chosen) and then about another axis perpendicular to the first and so on. In other words, the sphere can be thought of in the abstract as possessing any number of modes of rotational change *which leave its essential nature unaltered.*

The relevance of this to physics is that the physicist is preoccupied with deriving equations from the various conserved properties and then seeking transformations of these equations (or *rotating* them) to reveal that they are merely aspects of some single underlying form. Think of the laws of nature as a perfect sphere: then think of the equations denoting unvarying properties of particle behaviour as merely different rotations of the sphere. But all the rotations leave the sphere a sphere.

Now where does this get us? A little puzzled perhaps? It seems a long way from protons hitting protons, doesn't it? Let's remind ourselves again of what this is all about. Remembering the usefulness of our billiards game, let's look at another analogy.

This time imagine an observer, who has never seen the game, watching and attempting to understand a game of chess. Here the observer is the theoretical physicist, the pieces are particles, and their *moves* are particle behaviour, which obeys yet to be discovered rules. Of course, one must not mistake the pieces for the game. The definition of a 'pawn' lies, not in its shape or size nor the material of which it is made, but in the moves it makes. It would seem to be the same with particles, except that to extend the analogy to the problem of understanding particle states, one must imagine a chess game played with invisible pieces operating in, say, twenty-eight dimensions (directions in space)!

If we continue, we can see that rules of chess could perhaps be deduced from observing many games and noting what was constant, i.e. what was conserved. Our games student, in trying to deduce the nature of the game, would of course be led to invent labels for the different moves which he observed: for instance, a piece which moved diagonally might be labelled a 'cleric'—thus a 'cleric' would be defined as only moving diagonally and only moving on the same coloured squares (i.e., colour is conserved). A piece which only moved forward would be called something else, and so on. And as these labels became verified across the community of students of the game, probabilities of combinations or sequences of moves could be predicted and the predictions tested. The point that's being pressed here is the degree to which physics is no longer concerned with the *object*, the particle, but only with its behaviour.

Before we come face to face with one of the symmetries of particle physics, we ought to remind ourselves of the facts of the real game that the physicist has been observing.

In the particle collisions we have looked at, the higher the energy, the more new particles are discovered: but the rules governing their behaviour are not arbitrary.

Four basic kinds of interaction have been identified. Throughout these interactions there run patterns. There is book-keeping (conservation) of energy and momentum; of electric charge; of baryon number; of lepton number; of strangeness; of isospin. There are symmetries which relate matter with anti-matter; left-handedness with right-handedness, and future with past. There are symmetries which restrict the dynamical processes to give relationships between mass and charge and between mass and spin.

What's important to us here is, not whether these names have any very precise meaning for us, but that several quantities do exist that are conserved, and that they can be integrated into symmetries that simplify the whole picture a little. In fact what physicists have found is that, through some fancy mathematics, the particles that take part in the strong interaction can have their first order symmetries reduced from about half a dozen to only three 'labels': charge, hypercharge and baryon number.

Now, if we dare, let's look at how a more sophisticated symmetry works in practice. (Keep in mind that this is only one of many, many different kinds of symmetry in particle physics.)

In knitting, elaborately complex forms can arise from a small number of simple operations using perhaps only one thread. Through theory and experiment physics has deduced that the complex structures and interactions of matter can also be understood as a small number of simple operations taking place in the quantum field

Consider these images:

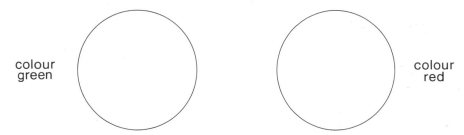

colour green

colour red

What do they show? A red disc and a green disc? Suppose these were a proton and a neutron: the only distinction between a proton and a neutron is that an 'electric eye' would see a difference between them: the proton is electrically positive, the neutron neutral. In all their strong interactions they behave alike. As we look at them we perceive that they are different, and we might be content to let them rest there in their diversity. But suppose we are mistaken to think of our proton model as a disc. The discs can be rotated this way, or this way, without changing anything. But what if we allow ourselves to extend our way of thinking about the particles into three dimensions? We might find that in fact spheres provide much more exact models than discs for what is known about their behaviour.

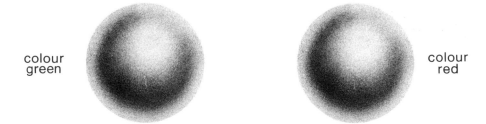

colour green

colour red

So we have imagined them to be spheres. What happens then if we rotate them?

colour green

colour red

colour red

colour green

We might see that in fact each sphere is half red and half green! And if we complete the rotation we have the same as before, in reverse. And this in fact corresponds to what is known about real protons and neutrons.

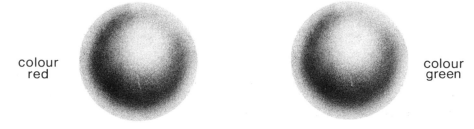

colour red

colour green

165

Through an imaginary rotation in what a physicist calls 'isospin space' we have revealed that a proton 'is' a neutron and vice versa. By the abstraction of 'think-ourselves-round-the-back-while-simultaneously-observing-the-front' we have been able to arrive at a much simpler view than before, i.e. that the two are essentially identical.

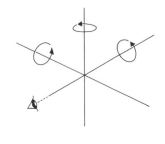

Up to now we have been thinking in three dimensional space, in which we could perform three imaginary rotations (diagram right), one of which (in line of sight to our eye) changed nothing. Now let us take a leap into space of more than three dimensions. Into say, eight-dimensional (eight-directional) space. Would a series of similar rotations in an eight-dimensional space show that many other particles are in fact really just other kinds of proton?

To see how this could be we have to make another big jump. Take it easy, don't panic.

If we represent the angle of rotation along the line of sight—away from you, by a dot: •
and the other two rotations by: ←
and by: →
we can combine them in a schematized way like this:

$$\leftarrow\cdot\rightarrow$$

Now, without calling it by its proper name, what we have here is a simplified representation of three operations: rotations in three-dimensional space if you like. And as we've seen this is a representation of the relationship of two nucleons, the proton and the neutron (and it also represents the relationship between the three different charge states of the sigma particle, for example). The isospin symmetry of the strong interactions, with its basic operations like the three rotations of a sphere which we symbolize $\leftarrow\cdot\rightarrow$, enables us to perceive the proton and neutron as just two different 'points of view', of the same basic particle, the nucleon. (And equally the Σ^+, Σ^o and Σ^- as three different points of view of the sigma.)

But we can go further and, by extending the isospin symmetry to what is called SU(3), show that the nucleon and the sigma are themselves different points of view of the same particle. To do this we need to be able to consider a different view-point from those which the rotations in the three-dimensional space allowed. SU(3) symmetry acts in a space with eight different directions.

The schematic representations of the rotations/operations in the eight dimensions of SU(3) leads us to add one more line-of-sight-type direction—another dot: •• and two more pairs of arrows thus:

together representing five rotations in five imaginary dimensions in addition to our first three. Together, they make this (diagram right):

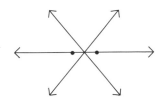

Note that each arrow schematizes an operation in space distinct from the others:

⟷   changes charge, but not hypercharge, and corresponds to I-spin

↖↓   changes hypercharge but *not* charge and corresponds to U-spin

↙↗   changes both charge and hypercharge

• •   one *measures* charge, the other *measures* hypercharge

Each label has a value of either 0 (no change) or +1 (positive change) or −1 (negative change). If we take our diagram through two further stages, we will perhaps see how this ingenious abstraction can show the strongly interacting particles to be indeed just variations on the same basic particle state. Take your time. Just let it wash over you.

First we should add the three labels to the basic diagram of our rotations:

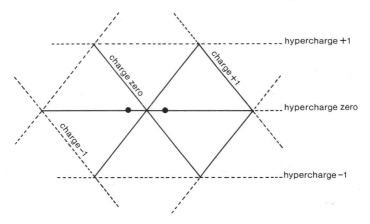

Now we can add the particles according to which of these values they have:

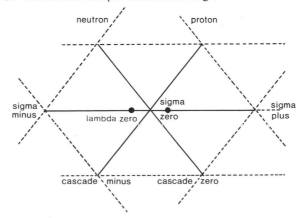

and now we can do the same thing for two other levels corresponding to baryon number 0 and −1 (the eight particles we've had so far all had baryon number +1). If we add to the diagram the appropriate strongly interacting particles according to their hypercharge, charge, and baryon number we get this:
In this three level array each particle is located opposite its own anti-particle.

Now we can perhaps see the way that, through their symmetry, the rotations relate together many apparently dissimilar particles, showing that they are all really versions, excited states if you like, of the proton (below right).

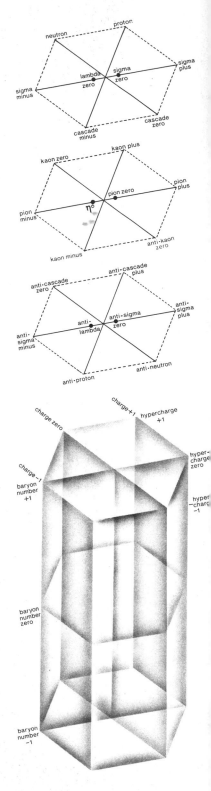

So what has been the point of all this? Just to show that, mathematically, a series of rotations in imaginary spaces can demonstrate that all of the particles in the above array are in fact just variations of the same particle state. Other more comprehensive symmetries have been proposed, which act in spaces of even higher dimensions, and relate still further properties of still more particles to one another.

What we've been trying to gain here is an insight into the frontiers of physics. If it has seemed difficult then we might take comfort in the fact that it's difficult for the physicist too. What he is engaged on is nothing less than the delineation of what is ultimately Real—what we have spoken of earlier as the 'normal on the cosmic scale'. This urge to find a single force or a single particle or a single equation which could be seen to explain the diversity of Creation is almost a parallel, it seems, to the age old pursuit of the Holy Grail, and it has been described thus by eminent physicists themselves. For many physicists the parallel may remain a distant or even unacceptable one; but it seems more than reasonable to take the symmetries of particle properties at this level as signposts, pointing inescapably to a totality of symmetry, a totality of harmonious interrelationship of all matter however or wherever it may be found. It seems impossible that these symmetries could be accident—coincidence. Our universe, it would seem, has been created in the only possible way; indeed it would also seem that it is so complete and so perfect that it is the only one that could exist.

But why should we limit ourselves like this? Only one universe? Wouldn't believing this mean that we wouldn't see an alternative universe, even if it could exist? This would be missing the point, which is that though there may be undiscovered aspects of the universe (which could account for ESP and so on), the completeness and perfection lie in the apparent certainty that what can happen has already happened or will eventually certainly take place.

Physics has achieved the reconciliation in our minds of wave with particle, matter with energy, time with space and motion with stillness; the next frontier that seems to emerge, and which was implicit it seems in quantum theory from its inception, is consciousness itself. If everything is totally and harmoniously related to everything else, will not experiment have to bridge the gap between 'the knower' and the 'known'? In other words, if particle symmetries do indeed point to a complete and whole unity of relationship, why don't we *experience* this unity?

168

11. Ignorance and Bliss

Science is by far the most powerful and universal system of knowledge in the world. It shows signs of having perhaps reached a certain maturity (even if measured only by the numbers of scientists and the money it consumes). The detail of established theories about the world may be disputed and in some areas there is very little agreement, but the basic facts of a scientific education are everywhere the same.

What is being suggested here is that science and especially physics neglects to take itself seriously enough—it fails to jump to its own conclusions. What are these conclusions? Can there be any conclusions?

For funding reasons, if no other, physicists resist any suggestion that physics has any 'conclusions'—that it is in any way complete. And it should be evident by now that any finality in the enquiries of high energy physics is unlikely (though as chemistry has shown there can come a point when enough is known for it to change from basic science into technology). Nevertheless it seems possible that out of all the physics we have looked at in the preceding chapters one thing emerges—the unity of the cosmos. Unity *is* the normal on the cosmic scale.

It is not suggested that physics *proves* the unity is there, but that the accumulated experiment, data and theory amount to a signpost. A direction sign. And the direction being pointed out is universal unity.

What do we mean by unity? It sounds comforting and satisfyingly grandiose as a conclusion for any kind of enquiry. But is it just a thought, like other thoughts? As the word is used here, it expresses the concept that all the diverse events and appearances in the phenomenal universe are totally interconnected and coherent, and that they arise from, they are manifestations of, a constant unchanging interior simplicity. The key word is total. For totally interconnected read—unified—union—a unity. As we have seen there are deep scientific intuitions of this in the conservation laws of physics, in relativity theory and in the symmetries of particle interactions.

The common factor in all these is constancy, invariance in the presence of change. This is the overall conclusion of physics: reality is that which is changeless. This is not a new idea.

> . . . that alone is real which exists unchanged and without intermission
> . . . continuity of existence without change is the test of reality
> . . . changelessness is an inalienable quality of the Real
> . . . the unreal has no being; the real never ceases to be
> *Ramana Maharishi*

Ultimate reality has been defined as that which never changes. Opposed to it is the unreal, which is ever-changing; for clearly that which always changes has no substance, no real existence
Maharishi Mahesh Yogi

Opposite top: We experience our ignorance of the unity of the cosmos as stress. Because they suppress the symptoms and do not touch the cause of the tension, most of our methods of reducing stress have a rather hollow quality. A cigarette advertisement blows a perfect smoke ring into a New York street
Below: the amount of stress we are experiencing determines the quality of our attention. It also determines the quality of our thinking. Very high levels of stress can so weaken our capacity to think that thought itself becomes just one more commodity in the market place

171

The surface of water may
assume a myriad of different
forms, each of which eventually
subsides, leaving the body of the
water unchanged. The Eastern
tradition teaches that however
attractive and real they seem we
should not confuse the
superficial ripples of movement
of thoughts, feelings and
sensation, with the deeper
reality of the mind itself

These are quotations from exponents of the traditional Eastern philosophies. There has been an insistence among Hinduism, Taoism and Budhism for many centuries that Reality is that which is unchanging and that the unity of the cosmos is the central fact of our existence. The particular importance of the scientific revelation of unity for both disciplines is that the evidence for this state that is being pointed to by physics has been arrived at through practical experiment, and has survived despite many restraints and checks intended to ensure that it is not just some primitive mental projection.

The importance of this basis of practicality has not been lost among those who carry on the traditional Eastern teachings, which are, despite their public image, also intensely practical.

When we investigate the invisible mechanics of nature, we find that everything in the universe is directly connected with everything else. Everything is constantly being influenced by everything else. No wave of the ocean is independent of any other. Each certainly has its individuality but it is not isolated from the influence of other waves. Every wave has its own course to follow, but this course is dependent on that of every other wave. The life of any individual is a wave on the ocean of cosmic life, where every wave constantly influences the course of every other.

According to the findings of modern physics, all matter has only phenomenal existence, and is in reality formless energy. Both in its previous state and its present obvious form, matter is nothing but pure energy, and on dissolution of the present form it will remain the same energy. Similarly, the present phenomenal phase of existence is seen to have no permanent significance . . .
Maharishi Mahesh Yogi

In fact a close contact with uncorrupted Eastern teachings quickly confirms that they too are based on practical experience but subjective inner experience. And because this experience is apparently unverifiable, their teachings have been until now rejected by most of us as belief or opinion. This now seems not to be so. They now appear as two apparently complementary paths to the same knowledge. Paradoxically it is science, which in its pursuit of truth has added so much to our sense of separateness, that has now come full circle to stand shoulder to shoulder with the findings of the Eastern sciences of Being. Make no mistake this juxtaposition is not a cultural pun (nor is it being suggested that science now equals religion).

The conclusion is inescapable: we must recognize that the fragile subtleties of high energy physics do have a bearing on our daily lives. Physics tells us that the universe we have is the only one that can exist—that it has been created in the only possible way. That's a physicists idea. Think about it, see how it relates to another Eastern insight:

. . . being eternal, Reality is all-pervading: being all-pervading, it is stable: being stable, it is immovable; being immovable, it is ever the same.
Baghavad Geeta

Does it also come as a surprise that physics has a message? Perhaps not.

If we have fallen into a state where we feel confused or frustrated or irritable then again we are likely to be experiencing stress. Stress is the experience of separatedness without the experience of unity. It would seem that the full value of life can only be lived when the separatedness and the unity are simultaneously experienced. So that for instance plants are seen and enjoyed for their individuality but also as eruptions of the earth towards the light, literally the soil unfolding in growth, and we in turn are an unfolding of soil through the medium of plant life

After all we expect the experience of a painting or a poem or music to be benign, why should we really slip into the belief (for that is what it is) that science has no content? That the medium of science is some kind of cold dead mountain of facts. Of course there will always be those who will defend this view, because for them to reject it would mean dissolving the contradiction between their work, which if they are physicists speaks to them of unity, and their personal ego-centric experience of the stresses of marriage and the bickerings of institutional life, which seems to speak to them of separatedness and aggravation.

Can we get over our surprise that high-energy physics could ever have any bearing whatever on the conduct of our daily lives? If so we might find it profitable to look at our lives, and see what if anything may be the consequences of the insistent pointing of physics and Eastern philosophy to the unity of the cosmos.

Can there be any consequences? Must we consign these concerns to some mental compartment reserved for interesting but impractical mind fluff? Mental activity to be taken out, dusted and worn at parties like a dowager's diamond brooch? Or is it something here that eats and lives in the nitty-gritty of daily life?

What does all this that we have been reading mean for the vigorous man and woman of action? What are the questions that come to mind? Why don't we experience the unity? If everything is perfect why do I need to even think about it?

If everything is perfect then of course you won't be doing anything about it. There's nothing here that you need. But only you can know whether your steady daily experience is that of the perfection of Reality. Why don't we experience the unity? This is where it seems the Eastern science of Being has something to offer.

Think for a moment of consciousness as a mirror. From the moment we are born to the moment we die, our consciousness is filled with perceptions of the outside world, together with the thoughts and feelings that we derive from those experiences. So like a mirror it is continually filled with changing images, and because of this it never experiences its own nature.
Maharishi Mahesh Yogi

Since the ignorant and the simple-minded, not knowing that the world is what is seen of Mind itself, cling to the multitudinousness of external objects, cling to notions of being and non-being, oneness and otherness . . . they are addicted to false imaginings . . . it is like a mirage in which the springs are seen as if they are real. They are imagined so by the animals who, thirsty from the heat of the season, would run after them. No knowing that the springs are their own mental hallucinations, the animals do not realize there are no such springs. In the same way . . . the ignorant and simple-minded . . . with their minds burning with the fire of greed, anger and folly; delighted in a world of multitudinous forms; with their thoughts saturated with the ideas of birth, destruction and subsistence; fall into the way of grasping at one-ness and otherness, being and non-being.
Lankaratara Sutra

Consider a field of grass in the early morning. On each blade of grass is a drop of dew, and standing at the edge of this field is a man who is very thirsty. And what does he do? He goes from one blade of grass to the next supping up the little drops of water. He goes from one blade of grass to the next not because of the fascination of the next drop but because the last drop had not satisfied his thirst. That is why our attention wanders in the relative field: because the last drop had not satisfied the thirst of the mind for joy. Now, tell that man there is an ocean close at hand, he just turns to it and drinks his fill. He is no longer thirsty.
Maharishi Mahesh Yogi

Beloved and Lover implies separation. And separation creates longing; and longing causes search. And the wider and more intense the search, the greater the separation and the more terrible the longing.

When longing is most intense, separation is complete and the purpose of separation, which was that love might experience itself as Lover and Belovêd, is fulfilled; and union follows. And when union is attained, the Lover knows that he himself was all along the Beloved whom he loved and desired union with and that all the impossible situations that he overcame were obstacles which he himself had placed in the path himself.

To attain union is so impossibly difficult because it is impossible to become what you already are! Union is nothing other than knowledge of oneself as the Only One.
Meher Baba

This is the message of Eastern science. Does it surprise you? Did the message in the physics pointing in the same direction surprise you?

If we are surprised by this message in physics, this in itself perhaps tells us that despite the expensive and extensive educations that some of us have had we have been conducting our lives somewhat in ignorance. We live in ignorance of the single most important fact of life, its unity—unity of nature; ignorance of the unity of ourselves, ignorance of the sheer scale of the unity of the whole cosmos.

The barometer of our ignorance is in the level of stress in us and around us. Each tightening of the neck muscles or clenching of the fists or setting of the teeth, each headache, is evidence of acting contrary to the natural flow of events, of acting contrary to the unity of the cosmos. Think about it, test it. Could you be more relaxed than you are as you read this? Are you smoking? If so why?

The more we experience feelings of suffering, the more we may be sure that we are trying to bend the inexorable purposeless flow of nature to fit our mental image of what we would prefer to be real. In our ignorance we often even think that suffering and stress are indicators that we are alive. That eczema and stammering and heart attacks and migraines are the price that has to be paid for an interesting life. Whereas they may only be nature's way of telling us that her cosmic law enforcement officers are aware of our wrongdoing.

Because the unity is *there* whether we like to be interested or conscious of it or not. It affects us constantly. Look at the evidence in the commonplace of our language. We speak of events going 'against the grain' we 'feel shattered', 'heart-broken'; things are 'neither here nor there'; we speak of 'working against the clock', of friends 'going to pieces' under strain, of couples 'breaking up', of feeling 'up against it', or 'all shook up', at 'cross purposes', 'dead against', 'all knotted up'. All these speak of our experience of the results of defying the fundamental fact of nature.

But we are also aware, if we let ourselves be, of the truth of the situation. We 'pause to collect ourselves'; we speak of 'picking up the pieces' of a bad piece of life; we tell friends to 'pull themselves together'; we 're-member' what happened last week or last year; we 'make up our minds'; without clothes we are 'in the altogether'; after a quarrel when we 'come round' to see another's 'point of view' (his sense of feeling separate) we 'make up'.

And also we are aware that it is a matter of consciousness, even if the word itself may frighten us a bit. This seems to be recognized when we speak about 'going out of our minds', 'taking leave of our senses', 'do you mind?', 'mind the traffic'; when we are angry we 'give people a piece of our minds', and when we feel particularly separate from the people we find ourselves with we are 'ill at ease'. It is only a little step from 'ill at ease' to understanding that even illness may have its roots in our defiance of nature; we suffer from disease. Next time you're ill, remember that the word really means 'dis-ease'.

It is often suggested that enlightenment, or coming to some experience of the cosmic unity, involves waking up into a state of super awareness compared with which our normal state is one of paralysis or inability to act. This (normal) state is often pictured, as here in the Doré engraving (below), as enchanted sleep

We know what's good for us too. We celebrate our delight in the spontaneous response to the natural flow of events when we talk about 'the moment of truth'; of acting 'on the spur of the moment', in 'feeling groovy', or 'being all of one mind'.

It seems that true and lasting happiness comes only from total and unremitting immersion in the flow of the unfolding universe. If we feel disturbed, agitated, angry, disappointed, or confused we may be sure that at that instant our ignorance is at work; that we are ignoring the unity of which we are inescapably a part.

It all depends on what we put at the centre of our lives, around what we centre our existence, on what we are prepared to acknowledge.

As Aldous Huxley has noted there are three kinds of things that we commonly put at the centre of our life: gadgets, political power and morality.

Its only too obvious that most of us have taken the gadget to heart and few people in the developed nations are free from the financial burdens of paying for and maintaining a whole catalogue of lawn mowers, cars, bicycles, fridges, washing machines, etc. And a large part of the national purse is in debt to the appetite for objects and the convenience they promise. But in our ignorance of the unity of all things and all life we forget that any local gain is necessarily paid for by some loss elsewhere; and so we mutilate ourselves in a host of ways as each new convenience that we add to our collection slowly and imperceptibly amputates an equivalent portion of our consciousness. We journey huge distances but don't *travel*. We extend our eyes through TV and amputate *conversation*. We record and transmit a staggering variety of sounds yet amputate real *listening*, and so on.

If we put political power for ourselves, or some candidate of our choice, or some organization, at the centre of our existence we merely exchange mechanisms—for dishwasher, read corporation or political organization. We move on from imposing mechanical solutions to our personal problems 'if only I had another car . . . I would be happy' to similar solutions to our social problems 'if only everybody was communist or republican then the class struggle would be resolved and happiness would result'. Just as we seek to impose the dishwasher as a solution on the problem of dirty dishes so we tend to try to impose the political solution on our social problems.

The third idolatry is that of the moral guardian. The protector from. The would-be film and book censor. Those who above all are dedicated to 'maintaining standards'; the organizations which try to protect us from ourselves. Each may be in itself necessary and appropriate but not as the total focus of our existence.

It is not being suggested that a hair dryer, or celebrities, or protectors of public morals, have negative attributes; that politics is unnecessary or that there should not be some limitation on the marketing of sex: just that if our attentive consciousness remains at a primitive level and if we centre our existence round *anything* other than the fact of the unity of the universe, then there will be a corresponding amputation of our consciousness. This is not the place to spell them out, these amputations; others have catalogued them only too well.

Overleaf: why has this memorial commemorating US Navy deaths in World War 2 unfolded in a form that so closely resembles the buildings that house the living around it?

Science and its branch of particle physics are not being depicted here as some addition to the time-table of global religious observance. We are merely posing the question: 'Why does physics not jump to its own conclusions?'

It's here that, as science converges with the findings at the root of Eastern philosophy, East and West may have the possibility of fortifying each other. Science has moved in its objective study of the material world from diversity to simplicity, and now points further, from simplicity to unity. The Eastern philosophy, or as we might think of it 'the science of being', takes the individual person from the diversity of personal subjective experience to the unity of experience. Both agree simply and exactly on what is real: 'Reality is that which is changeless.' 'That which is constant and unchanging in the flux of the phenomenal world—that alone is real.' Or we might prefer to say, Real. Some acquaintance with both disciplines leads to the simple view that if we look at what they both believe in, the apparently huge cultural gap between the committee charmers of big science and the man-charmers from the snake-charming countries is surprisingly small. So small that a merger of interests would not be surprising, and may indeed be far advanced already.

But beware of assuming that the merger would be total; one swallow does not make a Spring. Eastern philosophy is only concerned with *the experience* of the unity we have spoken of. Science presently holds itself to be concerned with *the description* of the world, and it requires only a moment's thought to realize that a description of unity is logically impossible. Description is intrinsically an action that can take place only through the separation of observer and observed. This is a barrier that has been recognized in particle physics for many years now. And one which is enshrined (if that's the word!) in quantum mechanics. Here are the views on it of three scientists:

The objective world simply *is*, it does not *happen*.
A. Weyl

To someone who could grasp the universe from a unified standpoint, the entire creation would appear as a unique truth and necessity.
J. D'Alembert

There lies the frontier, still almost as impassable for us as it was for Descartes . . . Brain and spirit are ideas no more synonymous today than in the seventeenth century. Objective analysis obliges us to see that this seeming duality within us is an illusion; but an illusion so deeply rooted in our being, that it would be vain to hope ever to dissipate it in the immediate awareness of subjectivity, or to learn to live emotionally or morally without it. And, besides why should one have to? What doubt can there be of the presence of the spirit within us? To give up the illusion that sees in it an immaterial 'substance' is not to deny the existence of the soul, but on the contrary to begin to recognize the complexity, the richness, the unfathomable depth of the genetic and cultural heritage and of the personal experience, conscious or not, which together make up this being of ours, unique and irrefutable witness to itself.
Jaques Monod

This is perhaps an ultimate frontier for physics. In the face of it, the pursuit

Opposite: the scientific pursuit of knowledge has allowed us to grasp at, and apparently hold in our hands, the whole planet. We have had great success in harnessing and controlling the power of nature and this very success has become a commodity, something to wear or to amuse us, but masking from us the meaning within the knowledge we have gained

of ever finer divisions, seeking the laws of illusion in the diversity of the universe may historically look like the actions of men fumbling in the dark, looking for evidence in the darkness for the existence of light. Whether this is too severe a view of where physics is at depends a lot on whether the scientist has at the centre of his life the continued existence of 'big science', or whether he has allowed his knowledge of the unity of the universe he inhabits to play some part in the direction of his life.

This is not the place for an equivalent consideration of the weaknesses of the oriental sciences of being; regrettably it is by their weakness that they tend to be known. Intuitively one feels that a cross-section of the relative excellence at any time of both sciences might reveal the same distribution of deep insight, delusions of grandeur and just plain ignorance.

The fact is that we are all in the front line of ignorance. Not only are we ignorant of the true nature of the universe, for which we might be excused; we are also ignorant of the true nature of ourselves.

We allow the idolatries of the gadget, the celebrity and morality (or immorality) to occupy our attention and reinforce the mental turmoil, which if we are honest we would admit is a large part of our experience. We try to be what we are not. We struggle. We suffer. But in truth we are none of these things. We may have open to us the possibility of an evolution of consciousness itself. As we become more aware of the inexorable flow of Nature we may see that on the cosmic scale there is no alternative, other than extinction, to the evolution of consciousness. It would seem that Western science as yet has no means of exploring or coming to a meeting with this situation. In the next chapter we tentatively offer some hints from the Eastern science which may encourage us to think that there may yet be ways of cleansing the doors of perception.

Opposite: Vajrasattva in union with the supreme wisdom Visvatara. A supreme Buddha form which personifies the state of being in which consciousness and knowledge are unified

Left: the explosive growth of scientific knowledge often seems to have benefited the personal lives of those people who contribute to it less than it might. The more knowledge in science and elsewhere has become separated from meaning and value, from the flow of growth and evolution, the more it has become another commodity. If the pursuit of knowledge through the methods of science were more mature, would not this gathering of eminent scientists for the opening of CERN's intersecting storage rings, also be simultaneously and unquestionably a gathering of *wise* men?

12. This is That

High-energy physics reveals that within the diversity of our surroundings there is simplicity. Unity. But in speaking of unity here and earlier we speak from separation. Here we are, you and I, *thinking of* unity. Wondering what the word means; perhaps weighing it. Trying to comprehend it. There's the trap. We are caught. The mental habit wins once more. We have happily assumed that unity was another concept for the mind to harvest. A nice thought for the mind to cosy up to. Cosmic unity. Wow! But we are kidding ourselves. Because unity means just that. Everything. Everything united in no-thingness. It *is*. Unity is beyond description, beyond language. It can be experienced, and people have spoken of the experience.

'There are moments when one feels free from one's own identification with human limitations and inadequacies. At such moments one imagines that one stands on some spot of a small planet, gazing in amazement at the cold yet profoundly moving beauty of the eternal, the unfathomable: life and death flow into one, and there is neither evolution nor destiny; only Being.'
Albert Einstein

'A kind of waking trance—this for lack of a better word—I have frequently had quite up from boyhood, when I have been quite alone. . . . All at once, as it were out of the intensity of the consciousness of individuality, individuality itself seemed to dissolve and fade away into boundless being, and this not a confused state but the clearest of the clear, the surest of the sure, utterly beyond words—where death was an almost laughable impossibility—the loss of personality (if so it were) seeming no extinction but the only true life.'
Alfred Tennyson

It is like the pool that becomes peaceful, quiet, any evening when there is no wind; when the mind is still that which is immeasurable comes into being.
Krishnamurti

That blessed mood.
In which the burden of the mystery,
In which the heavy and the weary weight
Of this unintelligible world
Is lightened;—that serene and blessed mood
In which the affections gently lead us on,—
Until, the breath of this corporeal frame
And even the motion of our human blood
Almost suspended, we are laid asleep
In body, and become a living soul
Wordsworth

Left: Om yantra. More than anything else, Om is the Eastern tradition's most ancient mantra—symbolising in sound the recognition that the fundamental reality of the cosmos is vibrational

Why do we not experience this unity? Did we know of its existence? If not, why not? What's stopping us right now? Five, four, three, two, one—unify! No change? There's no change because there is nothing we could do that affects the unity. The unity *is*. Remember? But by now perhaps we feel out on a limb. As if we are missing something. That's right! We are. We miss the unity because we

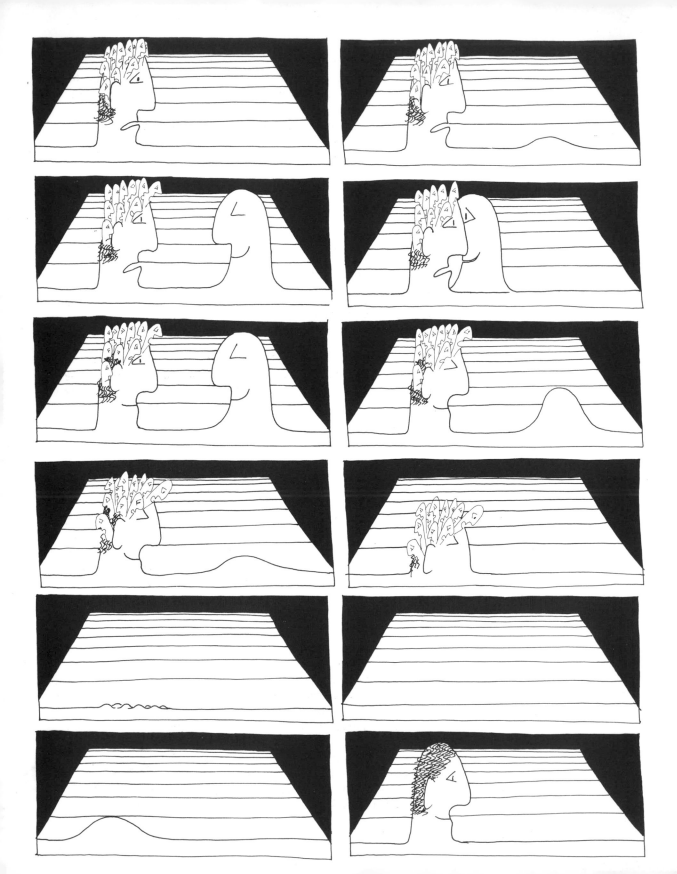

only use part of our minds. The surface. And our bodies we regard as separate machines, multi-purpose transport; and so we are less aware than we might be. We are apparently awake. But we are asleep to the reality of our surroundings.

If you try to make yourself content with the happiness of the pig, your suppressed potentialities will make you miserable. True happiness for human beings is possible only to those who develop their godlike potentialities to the utmost.
Bertrand Russell

How can we raise the level of our consciousness? How could we become Real-ized? Would it be good news?

'The atmosphere speaks for the quality of the person. The environment sings the glory of the individual. It always happens. A healthy, happy fulfilled man radiates happiness and fulfillment around him. The quality of the light speaks for the quality of the bulb. How powerful is the bulb? How clean is it? On that depends the radiation of light that comes from it. If the juice is sweet, the orange will radiate a sweet smell. It is the inner value that radiates outside.
Maharishi Mahesh Yogi

The experience of unity requires a method of work. A practical technique that takes us towards unity; that expands our consciousness until we are overtaken by Reality. Until then we must accept that what we know is very limited. One of the lines of the Rig Veda spells it out:

Knowledge is structured in consciousness

This is not a piece of mind fluff. Unless you fail to really meet it. What does it mean? It means you've grown up with a damp leaky bed-sitter of a mind. We both have. We didn't know that on the other side of the door was a palace.

The age-old answer to this problem has been and remains some form of meditation.

The very mention of the word no doubt sends many of you reaching for a mental strait-jacket in which to restrain the author for introducing such a topic in a book dealing with science. Don't let this resistance obscure the fact that if you want to make experiments in, and gain experience of, consciousness, some form of meditation is essential.

The last of the recipes for practical action that are recommended here is indeed a technique of meditation.

Many people come to meditation out of desperation, seeking a way to bring order into their otherwise chaotic lives, and one way of coming towards meditation is to consider it as a relaxation technique, a way of dealing with the stresses of modern life. It is only too obvious that stress and anxiety are the commonplace accompaniment to twentieth-century life; overstimulation leads to over-excitation leading in turn to massive accumulations of tension. In Western countries as high a proportion as fifty per cent of men die from diseases related to high blood pressure, which in turn is commonly attributed to tension. Even sleep itself seems not to be enough any more, and for many people it is inaccessible without the tens of millions of prescriptions for sleeping pills given every year. Meditation provides ready access to a state of deep rest with corresponding benefits of increased efficiency and mental clarity in daily life.

The author's experience of meditation has been gained through TM, the technique taught by Maharishi Mahesh Yogi. TM is a modern development of an ancient Indian system of meditation and it is taught according to a very carefully devised system which ensures that the technique is communicated very precisely. Great emphasis is laid on the effortlessness of the technique. Effortlessly one increasingly becomes able to dip into a state of profound peace, a state without thoughts, yet a state where all thoughts are possible. Effortlessly one learns to give up struggle. Effortlessly stress evaporates.

It might be thought from this that TM is a way of escaping from the day-to-day grind of the world, but this is not at all the case. In fact the practice is judged to be proceeding well or not in an individual by the extent to which he notices increased effectiveness in his routine actions, greater mental clarity, a reduction in tension, increased energy and greater creativity. This is an important point because the technique does not require withdrawal from the world— quite the opposite. Properly practised, it depends on the alternation of deep rest and normal activity.

The technique may be defined as turning the attention inwards towards the subtler levels of a thought until the mind transcends the experience of the subtlest state of the thought and arrives at the source of thought. This expands the conscious mind and at the same time brings it in contact with the creative intelligence that gives rise to every thought.

When a wave of the ocean makes contact with deeper levels of the water, it becomes more powerful. Likewise, when the conscious mind expands to embrace deeper levels of thinking, the thought-waves become more powerful.

The expanded capacity of the conscious mind increases the power of the mind and results in added energy and intelligence. Man, who generally uses only a small portion of the total mind that he possesses, begins to make use of his full mental potential.

Maharishi Mahesh Yogi

While its efficacy and importance as a method of simply bringing rest to those who need it should not be underestimated, meditation in our present context may perhaps be most directly demystified through another analogy—once more a musical one. If we think of ourselves as a kind of musical instrument, we can consider meditation as a tuning technique; we can see that, sadly, in our normal state we may be somewhat out of tune. We are aware that really we are capable of much more remarkable 'music' than in fact we are able to produce, and at times we can produce nothing but discord. Our mind is often out of relation with our body, our needs are at odds with our desires, the demands of work and marriage are often in conflict. And the gap between what we know we are capable of and what we actually achieve may be large enough to give rise to a great deal of frustration. To pursue the analogy one last step, it is fruitless to expect the human instrument to be really effective until the work of tuning has been completed or is at least under way.

The practice of meditation is a method of tuning, a way of bringing body, mind, thoughts, feelings and sensations into a harmonious whole. It's a way of

Opposite: mandala of Kalachakra—'the wheel of time'. With the exception of that on p 196, all the following illustrations are images from the Eastern tradition which have been, or still are, used to facilitate awareness of cosmic unity. They are vehicles for making experiments in consciousness itself; for inducing the experience of enlightenment

192

bringing the light of consciousness ever closer to us until at last it shines on, and therefore from, us.

Try it. Meditation can only take you where you need to go because remember, you already have everything you need; you have nothing to lose except your illusions.

No more words from me, but a few more words of gentle encouragement, food for thought, food for action, food for growth, from those who Are or have Been.

Below: a Shri yantra representing the primary divisions of the cosmic energy
Right: the Sanskrit alphabet—the vibrating essence of the whole of reality is believed to be contained within the sounds represented by the Sanskrit alphabet which is therefore in itself an image of whole-ness

Overleaf left: Field Ion micrograph, a direct image of the structural patterns of a tungsten metal crystal
Overleaf right: stalactite cupola of the Chamber of the Two Sisters in the Alhambra Palace in Granada, Spain. In buildings like the Alhambra, the Islamic architects gave expression to their intuitive belief that matter

was insubstantial. Islamic buildings all bear many repetitions of the statement 'La Illaha Illalah' which translates as 'There is one divine unity'

Page 198: another version of the Shri yantra

Page 199: Tibetan Tanka—a meditation diagram

Page 200: Shiva dances in a ring of fire, symbolising the recurring natural cycle of creation and destruction
Following pages: further details from the Alhambra Palace

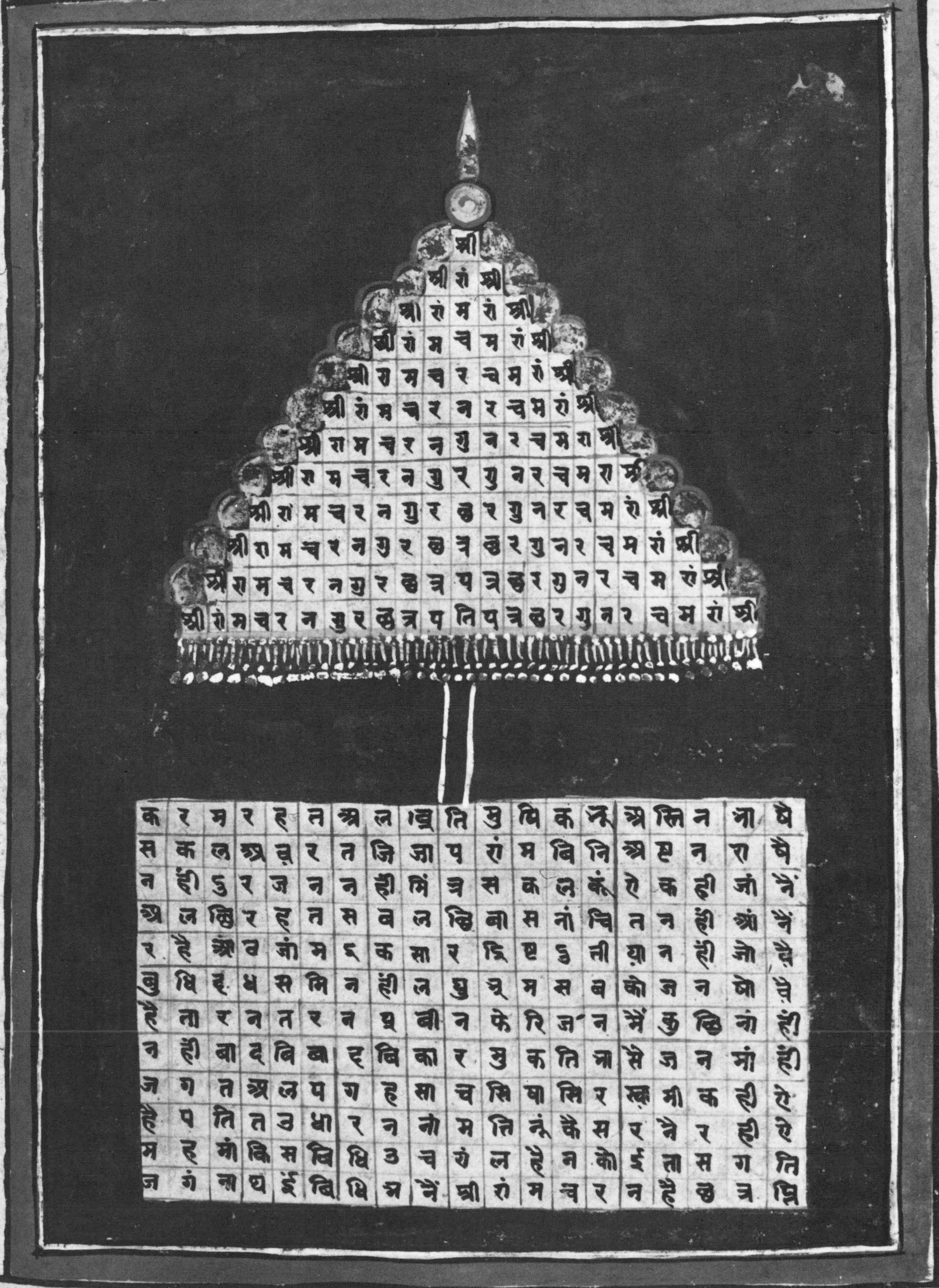
श्री रां म च र न गु र छ त्र प ति प त्र छ र गु न र च म रां श्री

That is perfect. This is perfect. Perfect comes from perfect.
Take perfect from perfect, the remainder is perfect.
May peace and peace and peace be everywhere.
Upanishads

I am the food, I am the food, I am the food; I am the eater
I am the eater, I am the eater; I am the link between, I am
the link between, I am the link between.
I am the first among the visible and the invisible, I
existed before the gods. I am the navel of immortality.
Who gives me protects me. I am food; who refuses to give me,
I eat as food.
I am this world and I eat this world. Who knows this, knows.
Upanishads

Beautiful

is the
unmea
ning
of(sil

ently)fal

ling(e
ver
yw
here)s

Now

E.E. Cummings

noone and a star stand,am to am

(life to life;breathing to breathing
flaming dream to dreaming flame)

united by perfect nothing:

millionary wherewhens distant,as
reckoned by the unimmortal mind,
these immeasurable mysteries
(human one;and one celestial)stand

soul to soul:freedom to freedom

till her utmost secrecies and his
(dreaming flame by flaming dream)
merge——at not imaginable which

instant born,a(who is neither each
both and)Self adventures deathlessness

E.E. Cummings

so many selves(so many fiends and gods
each greedier than every)is a man
(so easily one in another hides;
yet man can,being all,escape from none)

so huge a tumult is the simplest wish:
so pitiless a massacre the hope
most innocent (so deep's the mind of flesh
and so awake what waking calls asleep)

so never is most lonely man alone
(his briefest breathing lives some planet's year,
his longest life's a heartbeat of some sun;
his least unmotion roams the youngest star)

—how should a fool that calls him 'I' presume
to comprehend not numerable whom?

E.E. Cummings

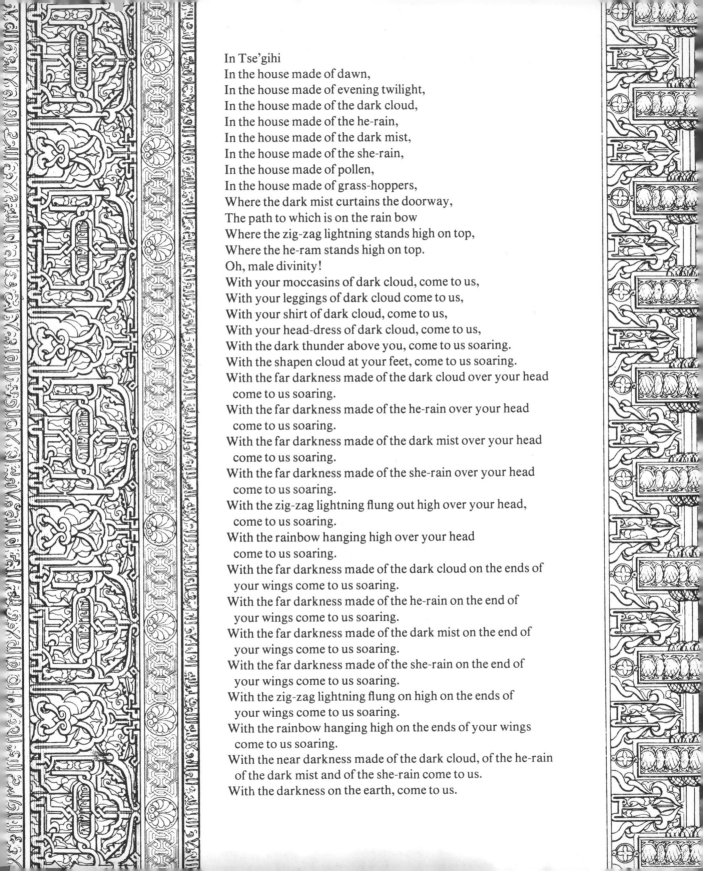

In Tse'gihi
In the house made of dawn,
In the house made of evening twilight,
In the house made of the dark cloud,
In the house made of the he-rain,
In the house made of the dark mist,
In the house made of the she-rain,
In the house made of pollen,
In the house made of grass-hoppers,
Where the dark mist curtains the doorway,
The path to which is on the rain bow
Where the zig-zag lightning stands high on top,
Where the he-ram stands high on top.
Oh, male divinity!
With your moccasins of dark cloud, come to us,
With your leggings of dark cloud come to us,
With your shirt of dark cloud, come to us,
With your head-dress of dark cloud, come to us,
With the dark thunder above you, come to us soaring.
With the shapen cloud at your feet, come to us soaring.
With the far darkness made of the dark cloud over your head
    come to us soaring.
With the far darkness made of the he-rain over your head
    come to us soaring.
With the far darkness made of the dark mist over your head
    come to us soaring.
With the far darkness made of the she-rain over your head
    come to us soaring.
With the zig-zag lightning flung out high over your head,
    come to us soaring.
With the rainbow hanging high over your head
    come to us soaring.
With the far darkness made of the dark cloud on the ends of
    your wings come to us soaring.
With the far darkness made of the he-rain on the end of
    your wings come to us soaring.
With the far darkness made of the dark mist on the end of
    your wings come to us soaring.
With the far darkness made of the she-rain on the end of
    your wings come to us soaring.
With the zig-zag lightning flung on high on the ends of
    your wings come to us soaring.
With the rainbow hanging high on the ends of your wings
    come to us soaring.
With the near darkness made of the dark cloud, of the he-rain
    of the dark mist and of the she-rain come to us.
With the darkness on the earth, come to us.

With these I wish the foam floating on the flowing water over
   the roots of the great corn.
I have made your sacrifice.
I have prepared a smoke for you.
My feet restore for me.
My limbs restore for me.
My mind restore for me
My voice restore for me
Today, take out your spell for me
Today, take away your spell for me
Away from me you have taken it
Far off from me it is taken
Far off you have done it
Happily I recover
Happily my interior becomes cool
Happily my eyes regain their power
Happily my head becomes cool
Happily my limbs regain their power
Happily I hear again
Happily for me the spell is taken off
Happily I walk
Impervious to pain, I walk
With lively feelings I walk
Happily abundant dark clouds I desire
Happily abundant dark mists I desire
Happily abundant passing showers I desire
Happily and abundance of vegetation I desire
Happily an abundance of pollen I desire
Happily abundant dew I desire
Happily may fair white corn, to the ends of the earth
  come with you
Happily may fair yellow corn, to the ends of the earth
  come with you
Happily may fair blue corn, to the ends of the earth
  come with you
Happily may fair corn of all kinds, to the ends of the earth
  come with you
Happily may fair plants of all kinds, to the ends of the earth
  come with you
Happily may fair goods of all kinds, to the ends of the earth
  come with you
Happily may fair jewels of all kinds, to the ends of the
  earth come with you
With these before you, happily may they come with you
With these behind you, happily may they come with you
With these below you happily may they come with you
With these above you happily may they come with you

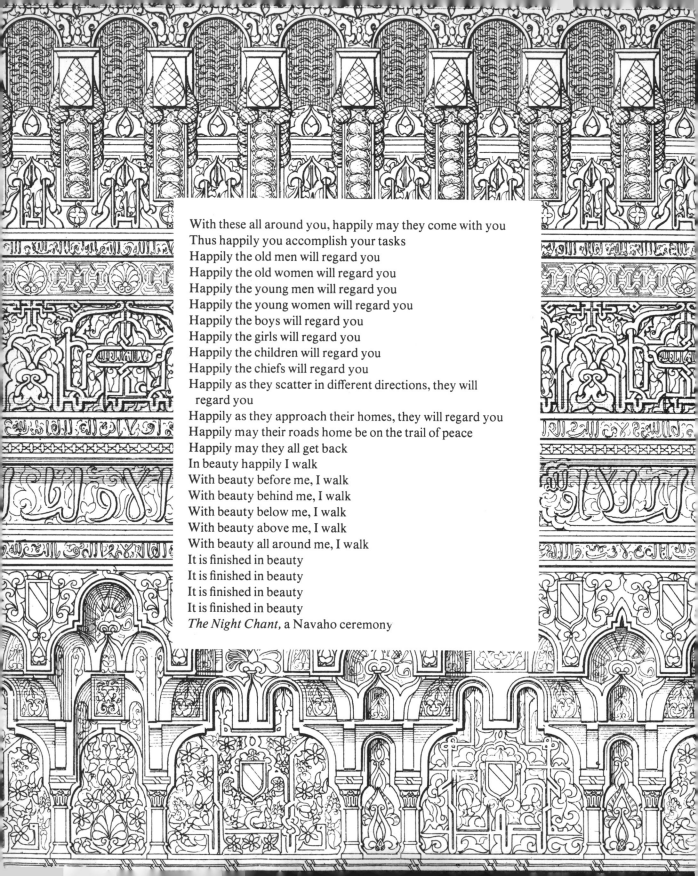

With these all around you, happily may they come with you
Thus happily you accomplish your tasks
Happily the old men will regard you
Happily the old women will regard you
Happily the young men will regard you
Happily the young women will regard you
Happily the boys will regard you
Happily the girls will regard you
Happily the children will regard you
Happily the chiefs will regard you
Happily as they scatter in different directions, they will
   regard you
Happily as they approach their homes, they will regard you
Happily may their roads home be on the trail of peace
Happily may they all get back
In beauty happily I walk
With beauty before me, I walk
With beauty behind me, I walk
With beauty below me, I walk
With beauty above me, I walk
With beauty all around me, I walk
It is finished in beauty
It is finished in beauty
It is finished in beauty
It is finished in beauty
The Night Chant, a Navaho ceremony

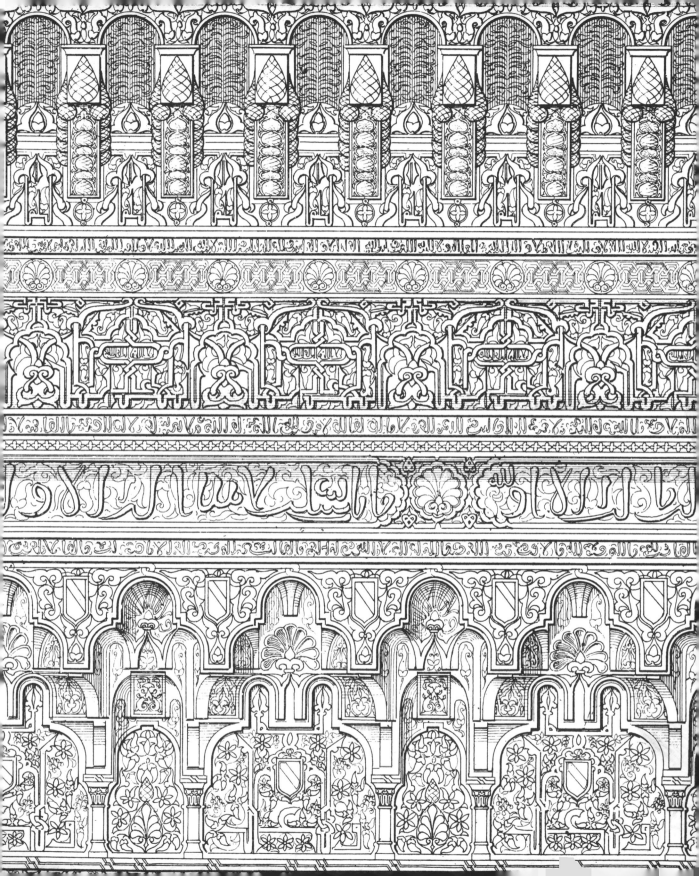

Bibliography

MAHARISHI MAHESH YOGI
On the Bhagavad-Gita A new translation and commentary
Chapters 1-6 Penguin Books 1969
MAHARISHI MAHESH YOGI
Transcendental Meditation The Science of Being and the Art
of Living New American Library 1968
ANTHONY CAMPBELL
Seven States of Consciousness Gollancz 1973
ANTHONY CAMPBELL
The Mechanics of Enlightenment Gollancz 1975
FRITZOF CAPRA
The Tao of Physics Wildwood House 1976
Sir ARTHUR EDDINGTON
The Nature of the Physical World Cambridge University Press 1948
JOHN M. IRVINE
The Basis of Modern Physics Oliver and Boyd 1967
ERNEST H. HUTTEN
The Ideas of Physics Oliver and Boyd 1967
RICHARD FEYNMAN
Lectures on Physics Addison-Wesley 1963

Picture Credits

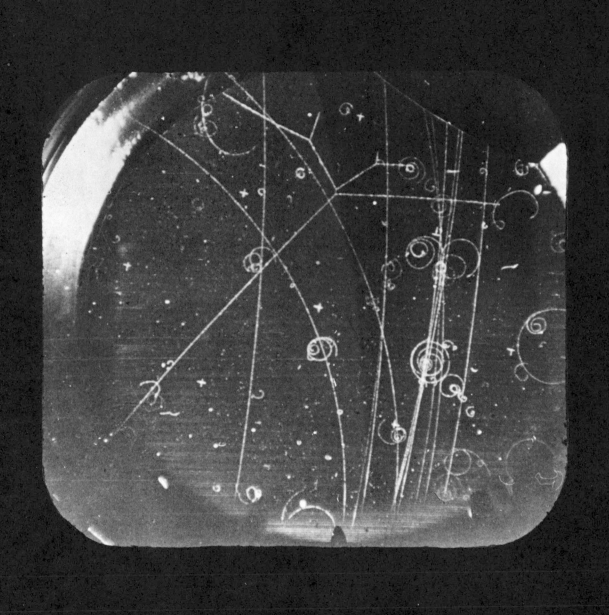